Finish Carpentry Illustrated

second edition

Finish Carpentry Illustrated

second edition

Elizabeth Williams
Robert Williams

TAB Books
Division of McGraw-Hill, Inc.
Blue Ridge Summit, PA 17294-0850

SECOND EDITION
FIRST PRINTING

© 1993 by **TAB Books**.
TAB Books is a division of McGraw-Hill, Inc.

Library of Congress Cataloging-in-Publication Data
Williams, Elizabeth, 1942–
     Finish carpentry illustrated / by Elizabeth and Robert Williams. -
- 2nd ed.
       p.   cm.
     Includes index.
     ISBN 0-8306-4410-5 (H) ISBN 0-8306-4411-3 (P)
     1. Finish carpentry—Amateurs' manuals.   I. Williams, Robert
Leonard, 1932–   .   II. Title.
TH5640.W55   1993
694'.6—dc20   93-12803
                                                                CIP

Acquisitions editor: Kimberly Tabor
Editorial team: Joanne M. Slike, Executive Editor
               Susan Wahlman, Supervising Editor
               Robert Burdette, Editor
Production team: Katherine G. Brown, Director of Production
                 Wendy L. Small, Layout
                 Susan E. Hansford, Typesetting
                 Tara Ernst, Proofreading
                 Nadine McFarland, Quality Control
Design team: Jaclyn J. Boone, Designer
             Brian Allison, Associate Designer
Cover design: Holberg Design, York, Pa.
Cover photograph: Thompson Photography, Baltimore, Md.   HT1

RECENTLY A retired North Carolina builder said that when he first started building, the customer could expect that 25 percent of his money would go for labor and the other 75 percent for materials . "Then a decade or two later, labor cost half and materials cost half." From the customer's viewpoint, the news has deteriorated since that time. "The last building job I did was the worst percent ratio I have seen. Materials cost 30 percent, and labor was 70 percent."

Although this trend may not be reflected nationally, it could be worse in some parts of the country. It is painful to many people, and their bank accounts, to realize that more than half of their building funds will be needed to pay for laborers who are often at best semiskilled carpenters.

The rest of the bad news is that if you should decide to build a house that will cost you $150,000 and you are fortunate enough to get such an unlikely interest rate as 4 percent, your payments will be $1300 per month for thirty years. This figure translates into a total payoff of $468,000—nearly half a million dollars. From a mathematical standpoint, you will be paying carpenters a total of $234,000. Although the carpenters will not receive nearly that much, you will be paying it in interest and principal when you repay the $150,000 you borrowed to build the house.

Such financial arrangements are staggering to most middle-class workers. A very small percentage of working Americans can afford a $1300 monthly house payment in addition to their other financial obligations.

There is a much better solution to the housing problem. You cannot find too many ways to reduce your materials cost; however, there are *some* ways to cut the cost of supplies. Whatever bargains you find, your materials will nevertheless cost you a huge portion of your building budget. When you have exhausted your money-saving efforts on materials, there is one other way to cut your costs dramatically, even by as much as 60 percent.

You can do the work yourself. TAB's *Rough Carpentry Illustrated*, *Second Edition*, (#4395) guided you through the initial efforts of construction—foundation walls, wall framing, roof framing, stairway construction, and the other major steps in rough carpentry.

This book is designed to help you complete the rough work you did in the early stages of construction and takes you through the final steps, including painting or staining the finished product and installing light switches.

Many people are intimidated at the thoughts of laying blocks or bricks, installing switches and receptacles, setting in bathroom fixtures, or hanging light fixtures. If they realized how simple these steps are, they would not continue to pay others $30 hourly to do the work.

If you are reasonably healthy and in at least fair physical condition, you can do all aspects of finish carpentry. You can save yourself thousands of dollars along the way and get an even better house for your efforts.

Consider finances for one more moment. Assume that you hire five carpenters at only $12 per hour each. In an eight-hour day, these workers will cost you almost $500. For a week's work, the total is $2400. You will have paid over $2000 for work that you can do in your spare time if you have some guidance. This book is intended to be your guide for doing the finish-carpentry work on your house.

If you have done the rough carpentry already, the remainder of the work should not be difficult at all. For the most part, you will be covering up the rough work with finish work.

Rough carpentry, as the term implies, is less than polished. While not necessarily crude, it does not often present an attractive appearance. *Finish carpentry* consists of what is seen by the world. The final stages of the work include hanging cabinets, laying or installing exterior wall covering and interior wall covering, building bookcases, hanging wallpaper, installing doorknobs, installing carpet, and adding all the finishing touches that make the difference between a makeshift house and a beautiful one.

Finish carpentry means completing your house as you wish your friends and neighbors to see it. It means adding the distinctive personal touches that no hired carpenter is likely to handle as you want them done. It means second-coating the wall because you know the surface is not up to your standards; it means countersinking nail heads and puttying the holes to give the door facing a flawless look.

Rough carpentry usually means using more equipment, doing more strenuous physical work, and exerting more strength and stamina. Finish carpentry requires more patience and ability to take endless pains if necessary to get the work done to your personal standards.

Pay a visit to a house nearing completion. Examine with a critical eye the quality of the work. In most instances the appearance is gratifying; if you look closer, you may begin to see the flaws. You can see brush strokes on the wood surfaces, evidence of drips or runs of paint down a wall, bent nails driven into the wood, absence of nails where they are required, uneven mortar joints in masonry work, ill-fitting wall switches or receptacle boxes, smudges of paint or dirt on a recently painted wall, and a host of other problems.

"I can walk into a house and spot the defects almost immediately," a former building contractor said. "It is appalling to see what levels of work are foisted off on the general public. People who know better are letting work that is unforgivably sloppy get past them, and the buyers often don't know any better and pay the price of high-quality construction for low-standard performance."

Think about economics. In the South, a cement-block mason charges about $1.50 for every block he lays. A hard-working mason can lay 500 blocks in a day, and his work is costing you $750 daily, in addition to the cost of the blocks and mortar.

The professional mason's work is usually very attractive. You will not be able to lay blocks nearly as fast and probably not as well as the mason can. With practice, you can lay a very acceptable block wall, and if you can lay only 200 blocks per day, you have saved yourself $300. Ask yourself how many jobs pay $300 daily. Remember the old adage "A penny saved is a penny earned." The money you save on the day's work will buy a great number of blocks or building materials.

Another adage insists that "Time is money." You may find that the house that would cost you $150,000 if you hired the work will cost you only half that much, even less, if you do the work yourself. If you can devote weekends and other spare time to building your house, you can complete it in less than a year. If you save $50,000 or more on the work, you have in essence earned an extra $50,000 that year. Better still, you have not paid interest on the money.

When we started to build our house, we learned from banks and contractors that the house we wanted to build would cost us about $150,000. With interest added, we would pay almost half a million dollars for the house over three decades.

We elected to do the work ourselves. We built the house for less than one-fifth of what the contractors wanted to charge. We saved over $120,000.

It was hard work, and it was time-consuming. We devoted every spare minute to the house, and it took us more than a year to complete it. When we were finished, we had 4300 square feet of living space with 13 large rooms or space for 25 rooms on three levels if we wanted to partition the larger rooms, some of which are 36 feet long.

Best of all, we paid for the building supplies as we used them, and when we were ready to move in, the house was paid for. We did not pay a cent in interest on the money we spent. If we count the total cost of the original loan plan, including interest, we saved more than $438,000 by doing the work ourselves. It would have taken us far longer then the year we devoted to building to earn that much money.

There are some points of advice that should be considered in addition to the financial factors. Some of the work can be done easily by one person. Most of it can be done by two people. If you have a helper, work out a time schedule so that two of you can be available for the heavier work.

Debate whether to buy or rent tools. You can rent an entire toolbox from outlet or retail stores, but you will be paying rent on many items you never use. You can pay for the tools in rental fees if you need to keep them very long. We suggest that you buy the tools you need as you need them. There is no reason to buy equipment that you will not need for another three or four months. Wait until you need the items, then make the purchase.

Rent large, very expensive tools that you will need for only a short time. Sanders are one example. Huge professional sanders are extremely expensive, and you can rent one for two or three days for far less than it would cost to buy one. You can also rent scaffolding, mortar mixers, and similar equipment that you will have little use for once the house is completed. A good rule of thumb is to estimate the number of days you think you will need the equipment, add at least one day for weather problems or other delays, and multiply the number by the cost of one day's rental.

Subtract the rental fees from the total cost of the equipment. If your rental will equal more than half the total cost, you might be well advised to go ahead and buy the item. You may be able to sell the equipment after your house is complete and recover at least half your initial cost. You could also rent it to others and recoup some of your investment.

Careful planning is essential for anything you rent or labor you hire. If you make arrangements to rent a mixer and the weather prevents your laying blocks that day, your rental cost is lost. The same is true if you hire labor to help you hang heavy cabinets and the cabinets do not arrive. You may have to pay at least part of a day's wages to the laborers. Take the weather into consideration when you are hiring workers or are starting a job that cannot be delayed once it is started. If you are pouring and finishing concrete, you cannot stop the job once you start it. If you are not finished by dark, you will need to rig lights and continue floating or finishing the poured floor. If heavy rains occur, your efforts and materials may be wasted.

Consider your physical strength and endurance. Do not rush into the work headlong but work yourself into shape by starting slowly. Do not mix three bags of mortar and then realize that you are too tired to complete the work. Mix smaller amounts, even if you must stop and mix several times.

Start with a set of standards and do not be sidetracked. When you are tired, you are more likely to accept substandard work that you would have rejected earlier. If you make a bad cut and decide to use the piece of wood anyhow, you will find that when you try to fill the crack with putty or a similar substance, the defect becomes glaring. Paint seems to accentuate defects in wall surfaces, and when the final decorating is done, you may be totally displeased with the finished product.

Develop patience. Do not leave the work until it meets your standards. When you leave work hastily and unsatisfactorily done, you will seldom return to it to improve it.

Maintain your confidence. You will have a number of people warn you that what you are attempting is impossible, impractical, or ill-advised. They will point out problems and defects in your work. Do not let their criticism affect your determination.

You will be discouraged at first by your slowness. Do not panic. As you work, your quality of work and speed will increase daily. Do not expect within a week to rival the skills, speed, and expertise of professionals who have done the work for years.

Maintain a steady work schedule. Do not let days slip away when you could be working. The longer it takes you to complete your house, the more rent you are paying, the more inconvenience you are experiencing, while you wait. Jobs that drag on too long become discouraging and tiresome. Stick to the work and get it done well in as short a time as possible.

This book offers as much meaningful advice as space permits. Drawings clarify some especially difficult aspects of your work.

The language is that of the layman, the beginner, the do-it-yourselfer. Highly technical terms are omitted unless they are essential to the the work. When unusual terms are used, they are clarified the first time they appear. In many cases, they are redefined when they appear in later phases of the work.

This book presupposes very little experience or knowledge concerning finish carpentry. It is simplified enough that the beginning do-it-yourselfer can follow instructions and complete all stages of the work.

References are often made to the need for building permits and building inspections. Please remain aware that if you do not have the necessary inspections made, your house cannot be completed, and power will not be turned on.

Do not demonstrate animosity toward the inspector. He or she is doing the job for one purpose: to help you to make your house better, safer, more enjoyable, and salable if you decide to put it on the market later.

Now it is time to begin work. Keep this book handy. Refer to it when you have questions. If your needs and plans differ from the ones given here, follow your own desires. Above all, be very careful. Serious injuries can negate savings. Good luck!

Completing foundation walls

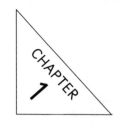

Tʜᴇ FOUNDATION wall of your house is probably composed of cement building blocks, mortar, and the support elements that strengthen the blocks and the bonding. When the rough carpentry is completed, the wall will be standing on top of the footings that were dug and poured earlier.

Now it is time to finish the walls. This may include only sealing and waterproofing, or it may consist of sealing, painting, installing drain tile and fill, and an assortment of lesser tasks.

Before you install drain tile, you may need to parget, or "parge," your outside foundation walls and waterproof them. This step is crucial, as you will learn if you omit it or do it ineffectively. As incredible as it seems, a heavy rainstorm that lasts for an entire day and night can funnel hundreds of gallons of water through leaky basement walls.

Parget work consists of covering the exterior of the cement blocks with mortar. Mix mortar as you would for any type of masonry work. Climb into the excavation space and place a mortarboard loaded with mortar at your feet.

If you do not have a mortarboard handy, make one by cutting two lengths of 2-x-4 about 2 feet long. Locate a square or rectangle of plywood about 2 feet on each side. Stand the 2-x-4s on edge and place the plywood section on top of them. Space the 2-x-4s so that they are under the edges of the plywood. Drive a series of 8-penny nails through the plywood and into the 2-x-4s. Seat the nails so that the edge of the trowel will not catch on them. The mortarboard is ready to use. Load it with mortar and place it conveniently in the excavation area.

Start at one corner and cut into the pile of mortar with the edge of the trowel. Hold the trowel upside down so that when you

Waterproofing foundation walls

load it, you will be able to spread the mortar across the cement blocks with one easy, sweeping motion.

Start low and sweep upward. As the mortar comes into contact with the blocks, apply enough pressure to push the mortar into the pores of the blocks. The mortar will adhere better if you spray or sprinkle water onto the cement blocks before you start to parget.

Let each strike slightly overlap the previous one. Continue along one wall until the entire cement-block surface is covered. Keep a bucket of water handy and dip your trowel into the water between strokes. The added moisture will thin the mortar slightly and make it more plastic. Smooth the mortar covering as you work. Always dip the trowel before smoothing.

This work moves rapidly and helps immeasurably to seal the blocks. Let the parget work dry before adding waterproofing to the wall.

You can buy any of several varieties of waterproofing material that you can apply with a paintbrush or any type of roller, trowel, or thick brush with stiff bristles. Apply the substance thickly and be sure to cover every area of the wall up to the ground level.

Most of the waterproofers are of a thick consistency and are tarlike. They are sticky and very messy, so you will need to protect other building materials from accidental spills or spattering. Clean brushes or rollers immediately after use, before the substance can dry. You can use any type of paint thinner or Varsol product to clean equipment.

Let the wall protection dry fully before you install the plastic covering. You can buy 100-foot rolls that will cover one entire wall with some left over. When you must join the plastic, lap it by 6 or 8 feet. Some builders like to tack or nail the plastic tightly against the wall. Others prefer to press the plastic into a still-tacky wall covering to hold the plastic in place. Still others tack the plastic to a thin strip of wood and tack the wood to the concrete blocks at a point above the ground line.

Installing drain tile

When the plastic is in place, you can install drain tile. At this point you have the wall itself, the parget cover, the waterproofing substance, and the plastic. Moisture will seep through the soil to the bottom of the wall, and the drain tile will carry the water away.

If you did not install drain tile during the rough-carpentry work, you should do this immediately. The purpose of this tile is to carry away from the house all the precipitation that pours off the roof or flows by gravity into your yard.

As much water as possible should be diverted from the house. Cement blocks are very porous, and eventually water will find its way into the basement or under the house—both undesirable conditions. Water under the house can create favorable breeding places for insects of all types, and the dampness can cause mildew and decay that can strike quickly, causing great damage. Water leaking into the basement can turn an otherwise excellent part of the house into a musty, damp, and almost useless place.

Assuming that you intend to include a basement (which we heartily recommend, partly because our basement saved our lives during a devastating storm that totally demolished our former house and partly because it is one of the most useful and economical areas in the entire house), you need to take several precautions against leakage.

Start by installing drain tile. You can buy this tile (with holes to allow surface water to enter the pipes and drain away from the building) in a variety of lengths. It can be cut easily with a saw or even a pocketknife and is very flexible so that you can bend it to conform to your excavation space handily.

Determine how many sections of drain tile you need. Plan to install it around the three sides of the house that are most susceptible to leakage. Measure the outside distance of the three sides and buy connector units so that you can create a single unbroken tile system. You will also need at least two elbows, or corner units.

To connect tile units, simply slip each end of the tile into the coupling. Some tile is manufactured so that one end slips into the end of the other. Examine the tile units to see how they should be connected. You can install elbows or corners the same way.

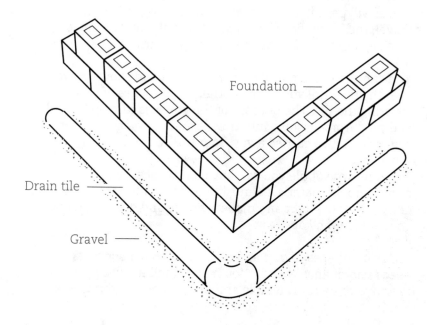

Foundation

Drain tile

Gravel

Some manufacturers suggest that you line the bottom of the excavation with at least 4 to 6 inches of gravel before you lay the drain tile. When you do so, rake the gravel so that it is more or less level. Do not leave humps that defeat the gravity-flow principles of the tiles. When the gravel is ready, start at the first corner of the house and lay the first tile. Connect the second tile and continue until you reach the first corner. Install the corner unit, or *elbow*, or *turn*, or *angle*, as the unit is alternately called, and proceed down the second side.

Follow this pattern of installation around the entire house on the three sides. At the end of the tile installation you will want to use additional lengths on each of the two narrow sides of the house. These added lengths run across the yard to carry the water away from the house as far as possible.

Adding gravel for drainage

Some builders like to dig a shallow trench that slants downhill slightly, then bury the tile in the trench so that it is not visible. You can also dig a square hole 2 to 3 feet deep and 18 to 24 inches square, then fill the hole with gravel.

Cover the gravel with dirt, then sow grass seed in the dirt. The gravel will permit drainage water to seep to the bottom of the gravel, then into the soil beneath. You will not have huge wet areas where the water has seeped to the surface.

You can then return to the drain tile area and shovel gravel over the tile until you have another 4 to 6 inches of gravel covering the tile. The purpose of the gravel is to keep loose dirt and mud from washing into the tile and clogging it so that water cannot be carried away from the house.

A second type of tile you can buy comes encased in nonabsorbent materials almost like the plastic used in packing crates. It is very light and maneuverable and does not require gravel under and above it.

Check both types of tiles and determine which one suits your needs and finances better. You may find that a little extra work can save you a considerable amount of money. From the other viewpoint, a little extra money can save you many hours of hard work. Choose the option that benefits you most overall.

Fill in the excavation when the drain tile is installed. If you have a chimney on one side of the house, you may need to install four bends or elbows to get the tile laid around the chimney footing.

You can hire someone with heavy equipment to push the dirt back into the excavation beside the wall, or you can do it yourself with a shovel and wheelbarrow. The decision depends upon your time, physical condition, and finances.

You can get someone with earth-moving equipment to handle the job, but you need to pay him by the hour for use of his equipment and his labor. Depending upon the part of the country, the costs can vary considerably. In 1992 in the South, earth-moving equipment and operator received about $50 per hour.

If you choose to save money by doing the work yourself, you can work out a schedule that permits you to haul 10 loads at the beginning of the day, 10 in the middle of the day, and 10 to finish the day. In this gradual way you will not overwork yourself, and within a few days the work will be done.

When the backfill is complete, you can turn your attention to sealing the remainder of the foundation walls. From the ground level to the top of the wall, you can brush or spray on a sealer that seals the pores in the blocks.

Apply one coating, let dry, then brush or spray another coating onto the wall. It won't hurt to add three or four coats of extra sealer. In a heavy storm with high winds, the driving rain can actually force its way between the mortar joints of blocks and flood or at least dampen and damage basements significantly.

When the outside work is done, go inside the basement and seal the walls there. Buy tubes of clear silicon caulking and a caulking gun. Seal the joint thoroughly where the wall meets the floor. If you buy clear caulking, the sealant will not be visible. If you choose one of the colored caulking compounds, you can always cover it with paint if you paint the walls.

When the floor joints are fully sealed all the way around the basement, apply the block sealer around the walls that are entirely or partly underground. Let the sealant dry, then add another coating. Again, you can add two or three coats. The cost of the sealer is not great, and the benefits are exceptional.

Normally you do not go to extremes on the various aspects of construction, but in this particular case you are not likely to overdo the protection against basement leakage. Continue to add to your precautions as long as there is a danger of water problems.

Much depends upon the terrain upon which your house is built. If drainage is not a problem, you can use less effort to prevent leakage. If the area around the house is not drained well away from the house, continue to waterproof the basement. Now is the time to do the work. Once the house is complete and the water heater, holding tank, and other equipment is installed, you will have a difficult time getting to the problem areas.

Once the basement is waterproofed inside and out, you are ready to complete the walls. You may wish to parget the entire inside of the basement, achieving a stuccolike effect that is attractive and lends itself to a variety of decorating motifs. You can simply parget the walls and leave them that way. You can also paint the walls when the mortar has dried. A third option for inside as well as outside is to hang rocks in the parget surface of the walls.

Hanging rock

To hang rocks, you must first obtain a good supply of high-quality rocks. These rocks are usually layered, the way shale formations occur, and they should be of a uniform surface and no more than 2 or 3 inches thick.

You can split the rocks along the grain. Like most wood, some rocks have a discernible grain that layers the rocks into an almost straight line. Hold the rock in one hand so that one edge is pointing upward. Use a masonry hammer to split the rock into two thin sheets or layers. A *masonry hammer* is designed much like an ordinary hammer except that the back part, where the claws would be in the typical carpenter's hammer, has a tapering bladelike design. Use this pointed end to split rocks. While you hold the rock, strike the edge in the center until it begins to split, as wood does. Tap lightly until the rock splits into two almost uniform halves.

To lay or hang the rocks, start by assorting them into piles made up of roughly the same size rocks. Try to keep the very large stones in one pile, middle-size ones in another, small ones in a third pile. If you have stones with straight edges, place these in a separate pile.

Mix the mortar to use on walls or chimneys. Mix about three parts of sand to one part portland cement, and then add water to achieve the proper consistency. It works somewhat better if you begin with 2 gallons of water, then add the sand. Mix sand and water thoroughly, then add the portland cement.

Do not attempt to hang rocks with mortar mix. You need the adhesive quality of cement.

When you have the proper mixture, continue to turn the mix with a shovel or rake it with a hoe until the sand and cement are mixed thoroughly. Work it until there is a degree of plasticity. The mortar should be thick enough to hold a shape when you cut it or mound it but thin enough to be spread on rocks.

You can start at the bottom or the top of a wall. You will not be laying rocks in the sense that the bottom ones support the top ones. You will be sticking each individual rock to the wall.

Use the larger and the thicker stones at the bottom. In this manner, if the wall is not perfectly straight, it will tend to slant toward the house, which is what you want. Do not let it slant away from the house.

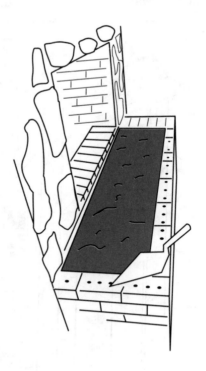

You can now "slush" the stone area. The mortar is the consistency of melting snow, and you can load a trowel with the mortar, then sling or throw it onto the wall. When the area is covered with mortar, hold the stone in one hand and push it with a quick shove against the wall. There should be moderate pressure against the stone to push it into the mortar.

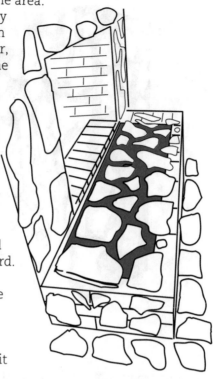

While you are holding the stone in place, twist it a third of a turn while pushing inward. This movement creates a vacuum effect that will cause the stone to adhere to the cement mixture. Maintain pressure for several seconds. When you release the stone, it should stick to the wall.

If the stone falls, reslush the area and hang the stone again. Hold it longer this time. When you release it, see if it sags slightly. If it does, apply pressure again and turn slightly more. Hold the rock until it sticks without sagging.

When the stones adhere, move up or down the wall and continue to slush and hang. Try for an interesting mixture of shapes or hues in rock surfaces.

Work until the entire wall area is covered. Use the stones with square edges in corners for a neater appearance. Do not use stone that is unusually thick or humpbacked.

When the wall is complete, mortar between the stone edges. You can use black mortar or traditional colored mortar. Do not try for a neat, trim effect.

Mortaring between stones

Load the trowel and sling the mortar into the cracks between the rocks. When the cracks are filled, add a little more mortar until it is beaded out past the surface of the stone. Use the point of the trowel to spread the mortar so that it laps over the edges of the stones. When you have done this, you will see that the covered edges give the effect of a perfect-fitting stonework.

Some people prefer to start at the top and work down because any spilled mortar will cling to the rocks below and will be very difficult to remove once it hardens. Even when you remove it while it is wet, some discoloration occurs.

Black mortar is likely to create problems, so be very careful that you do not allow any of the mortar to touch the surface of the rocks. You may never get them cleaned completely. If you choose to cover the foundation walls with bricks, you use a different technique altogether (see chapter 2).

Laying bricks & blocks

LAYING BRICKS and blocks is much easier in several ways than laying or sticking rock or stone. All the bricks or blocks will be of the same dimensions and therefore easier to handle for a straight, even wall. The problem with bricks and blocks is that you must maintain a very straight system of bonding and a straight course level. A *course* is one row of bricks or blocks, and courses that dip or rise in the middle have been laid poorly under most conditions.

To lay bricks or blocks, you will need a mortar pan or box, a trowel, a masonry hammer, a supply of good clean masonry sand, and a line level. You can also use a good supply of strong cord to use as guidelines.

Mixing mortar

Start by mixing mortar. Do not use sand from a creekbed or the side of a road. Such sand has lumps, pebbles, and various bits of foreign matter in it.

For large jobs, start with 5 gallons of water poured into the mortar box. Add 8 shovelfuls of sand to the water and mix thoroughly by pulling a hoe through it or turning it with a shovel. Add one 70-pound bag of mortar mix. Do not use portland cement for bricklaying. Continue to mix until the sand, mortar mix, and water are fully integrated.

Little by little, add 16 more shovelfuls of sand to the mixture. Mix completely. Do not stop mixing because the sand and mortar are wet. All the ingredients must be fully mixed into one plastic mass. The mixture should be thick enough to hold its shape when cut or mounded and to hold the weight of bricks or blocks. It must also be thin enough to be workable.

When mortar is ready, set up corners of the wall. You already have the block foundation wall in place and you are now ready to lay the brick wall to cover the foundation wall and to continue to the eaves of the house, if you plan to brick veneer the entire structure.

Clean off the footing so that you can lay a bed of mortar several feet in two directions. Do not leave dirt and other debris on the footing.

Laying bricks

Lay the mortar bed, then lay the first brick so that the end of the brick aligns with the corner perfectly. Push the brick gently down into the mortar so that the mortar is pushed up through the holes in the bricks.

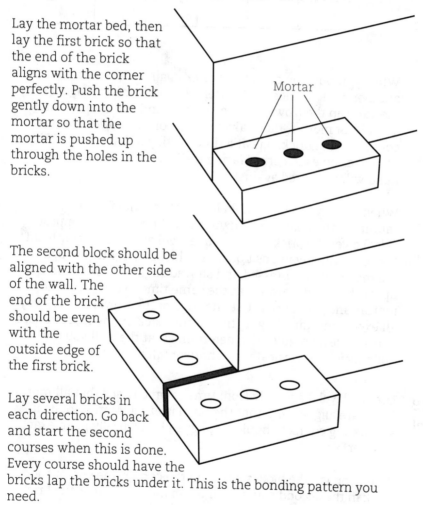

The second block should be aligned with the other side of the wall. The end of the brick should be even with the outside edge of the first brick.

Lay several bricks in each direction. Go back and start the second courses when this is done. Every course should have the bricks lap the bricks under it. This is the bonding pattern you need.

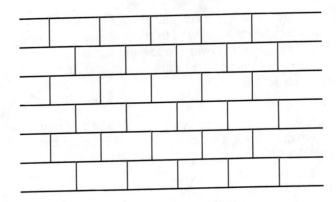

When you butt the end of one brick against the end or side of another, you will need to "butter" the brick you are laying with mortar. Dip the trowel into the mortar and scoop up a small amount of mortar, then rake the edge of the trowel over the corners of the brick, one at a time, until you have a small ridge of mortar on each corner facing the brick the brick in your hand will be butted against.

When you lay the brick, hold it across the middle with your thumb on the near side and your four fingers on the opposite side. Lower the brick to the course level and push it gently into the brick already in position. Push the brick tightly enough that the mortar is compressed and shoved slightly outward and upward. Push downward at the same time to compress the mortar under the brick. Use the edge of the trowel to cut away all excess mortar, slinging the mortar back on the mortarboard. Use your level regularly to determine that the wall you are building is true vertically and horizontally.

Using the line level

Use the level diagonally from time to time to see that all mortar joints are aligned and that the bond is exact. Mortar joints in alternating courses should be aligned horizontally in an exact manner.

When you lay the blocks out from the corner, the end of each block in the second course will reach halfway across the brick under it. Be sure that this bonding remains constant.

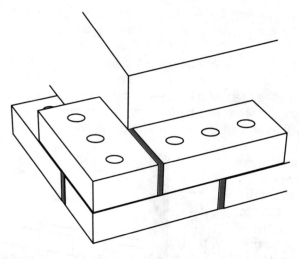

Keep building up the corner until it is waist-high. Then go to the next corner and repeat the pattern. This time run a line level from the first corner to the corresponding course on the other corner.

When you have laid the corner bricks and several bricks out in both directions, start the second course. Reach around the first corner and insert a nail in the mortar joint between the second and third courses. If you have a line block, use this instead of the nail and level line.

At the second corner insert a nail in the mortar joint between the second and third courses as you did at the first corner. Loop a line around one nail, pull the line taut, then loop it around the second nail. Hang the line level on the taut line. The bubble should be centered to show you that the two corners are exactly the same height.

If there is a slight variation, either thicken the mortar joint or compress it, depending on whether you need to raise or lower the line. If you do not start with level corners, you will have serious problems when you get to the higher courses.

Once you are satisfied with the level reading, locate the story pole you used in the rough carpentry phases of the house. If you don't have one, you can make one easily and

Using a story pole

inexpensively, using a 2-x-4 or a board to make the story pole. All you need is to mark the level for each course of bricks.

Bricks come in a wide range of sizes, as do cement blocks. If your bricks are 2½ inches thick, the mortar joint will make them 3 inches high. Mark off your story pole in 3-inch gradations (or whatever height your bricks plus the mortar joints will equal) all the way to the top of the pole.

As you work, lean the story pole against the wall and from time to time check the markings on the pole against the height of the various courses you are laying.

Checking the courses

Now you have to check points. You can use the line level to show you that the corners are the same height, and you can use the story pole to see that courses are as high as they are supposed to be. A third check is the guideline. Remove the line level and use the guideline to show you where the top of each brick in a course should come.

If any brick in the course is higher than the line, tap the brick down with the butt of the trowel handle. If a brick is too low, thicken the mortar bed under it and bring it up to the correct height.

When both corners are waist-high, start on the other two corners. Work exactly the same way, using line level, guideline, and story pole. If you keep checking, you can assure yourself that there is a perfect line level from corner A to corner B, then from corner B to corner C, and from corner C to corner D. If you do this, there is little danger that your courses will be out of line with the various corners.

If you do not make these checks from all four corners, you may find that when you reach the third or fourth corner later, you will have one brick that is only half as high as the one it is to bond with. Such an alignment is unacceptable, and you will have to go back and remove the bricks that are out of level.

When all four corners are waist-high and aligned correctly, you can begin to lay bricks between corners and bring the entire wall up to the corner levels. Choose the starting wall and lay a mortar bed the entire length of the footing between the two corners. Go back and lay bricks along the entire length of the mortar bed. Stop to cut away excess mortar at each brick. Fill in the course, then raise the guideline and use the line level and the story pole to see that you are keeping courses aligned as they should be.

If you are a beginner at bricklaying, you will be fairly slow at the beginning. For this reason you need to pause from time to time and "scratch out" the mortar joints. The purpose in scratching joints, or *jointing*, is to push the mortar back into the joint enough to cause it to adhere well to the bottom and top surfaces of the bricks. If the mortar does not adhere, your courses will not be stable and strong.

Jointing

Do not wait until the mortar has set up completely before you joint the work. Neither should you do the jointing while the mortar is so "green" or soft that you will scratch it completely out of the joint.

You can buy a jointer at a hardware store. You can also use a short length of 2-x-4 or a brick held at a three-quarter-degree angle. Turn the brick or 2-x-4 (no longer than 18 inches) so that one corner fits between the bricks at the joint. Push the jointer forward, starting at a corner, and continue to joint with short, steady, even strokes. Complete the entire series of courses horizontally, then joint them vertically.

Before you are finished with that part of the work, use a stiff brush or broom and sweep the excess mortar off the bricks. Once it hardens, it is difficult to remove.

Completing the walls

When all four walls are brought up to the level of the corners, continue to build the corners until they are at the limits of a good scaffolding height. This means that you should be able to work comfortably from the scaffold without stooping or reaching too high. The best level is from the waist to the chest of the mason. When you reach eye level, it is time to raise the scaffolding.

When you must work from a scaffold, rig the scaffold, if you have enough units to do so, from one corner to the next. In this manner you can complete entire courses without having to move the scaffolding, which is troublesome to take down and set up for a few minutes of work.

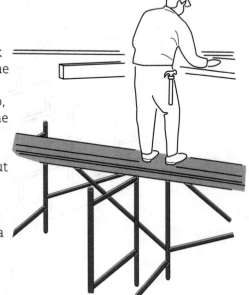

Complete the entire wall while you are working in that area. Then move the scaffold to the next area and complete the entire wall. The less you must move equipment, the more efficient your work schedule will be.

Installing lintels

When you reach windows and doorways you must lay bricks across the opening. You must install metal rods or lintels on which to lay the bricks.

A *lintel* is a long strip of very strong metal that stretches across the entire window or door opening. Lay a mortar bed up to the end of the course of bricks, then lay the lintel across the opening so that the ends of the lintel are embedded in the mortar.

When you are ready, lay a mortar bed across the lintel and lay bricks across its length. On the next course, lay a bed of mortar across the bricks, just as you would if you were following a regular course of bricks. Continue to lay bricks as before. The lintel will keep the course from sagging and interfering with the opening and closing of the door.

Cement blocks

If you decide that you want the entire wall to be built of cement blocks, you will find that there are many highly attractive and decorative blocks you can choose from. These blocks are as dressy and fashionable as bricks.

Lay blocks essentially the same way you do bricks. You must level across the block from side to side and from front to back. You must maintain a constant check on the vertical trueness of the wall and check on the bonding pattern as you work.

You will not have a wall frame as such when you make the exterior and interior walls from blocks. Start your corners as you would if you were using bricks. Build up corners, then lay the blocks in between.

To butter a block, stand it on end with the "ears" of the block upward. Scoop up mortar on the trowel and apply it to the edge of the block so that the mortar is embedded in the furrows of the ears. Turn the trowel sideways so that you can compress the mortar slightly and press it into the ears of the block.

With your mortar bed already laid, lift the block by grasping it at the thick section of the core separation. Use your other hand to guide the block into position. Set it as gently as you can into the bed of mortar and push it gently but firmly against the previous block. Cut away all excess mortar, tap down high corners, and level the block.

You may want to fill the blocks with some type of insulation as the wall rises. You can pour the insulation from bags into the blocks and fill them to the top. At the end of a workday, be sure to cover the block wall with sheets of plastic to keep the insulation from becoming wet if it rains while you are away from your work.

At the top of the wall you will need to fill in the blocks with solid mortar or concrete mix. When the walls are complete, you can parget or stucco the blocks for a distinctive look.

Joint or scratch out cement-block walls just as you did brick walls. Brush or sweep the walls at the end of the day's work so they will be neat and clean the next day.

If you choose to paint the blocks, you can seal them with a liquid available at most building-supply stores. The sealer penetrates into the pores of the blocks and makes them watertight and essentially airtight so that they will be impervious to high winds and driving rains.

Cleaning equipment

With all of your masonry work, cleaning up after the work is done is a vital part of the operation. Clean your trowel until it shines. A rusty trowel will not deliver mortar in one fluid sweep of your wrist. You will find that you must sling the mortar off the trowel or scrape it off, and the result is less than highly attractive work.

Wash out the mortar box daily after use. If you don't, it begins to accumulate mortar along the bottom and sides. If you use a mixer, clean it meticulously. We have seen mixers so clogged with dried mortar and concrete that there was not enough room to mix more than half a run of mortar.

Keep rakes, shovels, and hoes clean as well. If you allow mortar to accumulate on the tools, they will weigh twice as much as they should, and you will work much harder than necessary to get the mixing done.

When the masonry work is done, you are ready to pour the concrete floor for the basement if you have not done so already. Some people like to pour the basement floor and lay the blocks upon the concrete floor.

CHAPTER 3

Pouring concrete

POURING AND finishing concrete is one of the toughest jobs in house construction. So much depends upon the final work that you cannot afford to be careless, and once you have finished the work, you cannot go back and make minor changes.

Cutting costs
You can take several steps to reduce the work load and save yourself some money. Assuming that you plan to concrete the entire floor of the basement and the patio outside the basement door, you can begin at once to render the work area more accessible or more prepared.

Start by rigging temporary lights throughout the basement area where you will be working. Be sure that you can see a line-level reading anywhere you work.

Your first step is to begin digging out for the concrete. When the basement was excavated, you were left with a reasonably accurate floor space, but the heavy equipment is seldom likely to get the readings exact. You want 4 inches of concrete poured over the floor and patio. Measure from the ground level to the bottom of the floor joists. You may find that you have only 8 feet of clear space, which is normally enough for a ceiling, but this is not enough space for a usable basement. You need 4 inches for concrete and 4 inches for the gravel under the concrete; you also need 2 inches or more for ceiling cover and light fixtures. By the time you have taken away all this space, you will not have enough headroom for adults to move easily.

You need to dig out at least 6 or 8 more inches of dirt. Start in the center of the floor and using a shovel and mattock or pick, dig a hole 8 inches deep and 1 foot wide. Use the measuring tape to see how much space you now have to the bottom of the hole.

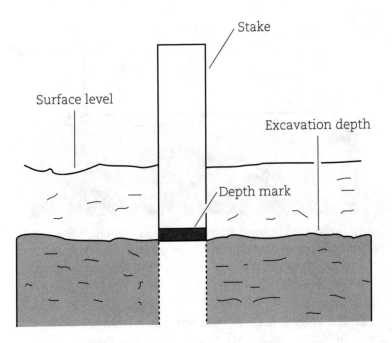

Stake

Surface level

Excavation depth

Depth mark

You now have 8 feet 8 inches of space, which is adequate if you don't require a great deal of headroom and you plan to hang light fixtures against the ceiling. Remember that ductwork for the heating system must be suspended from the ceiling later.

Drive a stake into the center of the hole. Measure down from the bottom of the floor joists 8 feet 8 inches. Mark the stake clearly at that point.

Begin to dig in a circle out from the stake. Shovel the dirt into a wheelbarrow and haul it to a place that needs filling. Keep moving outward from the stake.

When you are 5 or 6 feet away, attach cord and line level to the stake at the mark you made. Pull the cord from the stake to the limits of your digging. Hold the cord low enough that the line level shows a horizontal reading. If the cord cannot be held low enough, dig out another inch or two. You should be able to walk in a circle around the stake and keep the horizontal reading at all points.

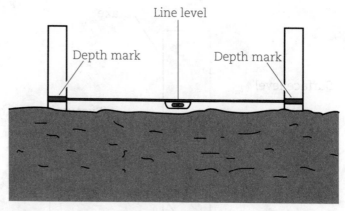

Line level

Depth mark   Depth mark

Continue to dig outward from the stake until you have reached the walls. This may take several hours or several days, depending upon the size of the basement, the time you can devote to the digging, and the difficulty of the terrain.

Marking basement-floor depth

When the digging is done, measure down from the bottom of the floor joists 8 feet. Mark this point on the stake you drove into the center of the basement. Next run the line-level cord

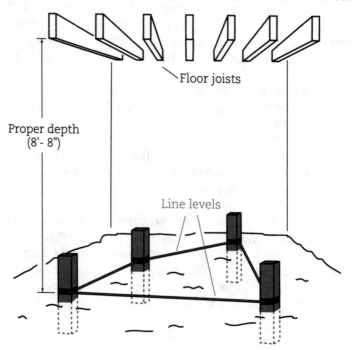

Floor joists

Proper depth (8'- 8")

Line levels

out to the basement wall at several points. Mark the level spot on the basement wall in the corner, 8 feet away, then 16, then 24, then 32, until you reach the end of the basement.

Move around the entire basement area and mark the walls. When all the marking is done, use a chalk line to strike a mark along all exterior walls to connect the points marked. This line is the top surface of the basement floor after it is poured.

When pouring is in progress, the concrete should come up to but not past the line you chalked. At the doorways into the room sectioned off by the bearing wall, mark the top level of the floor, then drive a stake in the center of that room. Measure down 8 feet and mark the stake. Use the line level to get the top points of the wall along the interior of the room, then chalk the lines as you did before.

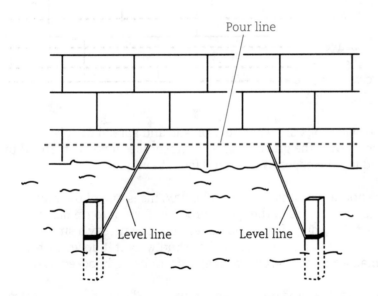

You now have all interior and exterior walls marked for the top of the concrete. The next step is to haul a 2-inch depth of sand into the basement. Dump the wheelbarrow loads of sand, then use the line level from a point marked at 8 feet 6 inches. Be sure that the sand is smooth and allows for headroom.

Mark another line at 8 feet 3 inches and begin to haul gravel into the area. Dump, spread, and smooth the gravel until it is at the height of the line level. Now you are ready for the plastic sheets to cover the gravel.

You now have room for 3 inches of concrete, but as the concrete is poured, its weight will cause the gravel to settle slightly, and you will have about 4 inches of concrete poured on top of the gravel.

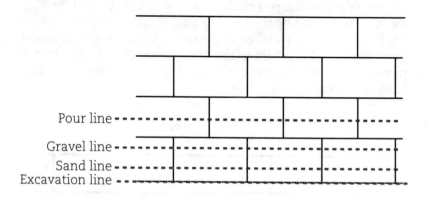

Buy wide rolls of heavy-duty plastic and spread the plastic sheets over the gravel. Use two thicknesses if you can afford to and lap the edges by at least 3 feet at every point.

Do not be careless with this plastic. This is one of the most important steps in the entire process. Try not to rip holes in the plastic at any point. If you do tear the plastic, lay a small sheet or square of plastic over the hole and weight it down with a small amount of gravel to keep it from being knocked aside.

When we poured our basement, we failed to patch three holes, and each time we had a prolonged heavy rain, small patches of moisture appeared on the basement floor. Several treatments with a sealer were required to eliminate the seepage.

When the plastic is in place, you are ready to have the concrete poured. Call the company and make arrangements for delivery.

If you want a drain in the basement, you need to take care of the base work before you add the sand, gravel, and plastic. You can buy the drain and pipe at a hardware or building-supply store. Install the drain at the proper level, just slightly lower than the stake and wall marks.

When the concrete is poured, the top will be even with the housing of the drain. Because of the slight incline from the walls to the drain, any accumulated moisture will cross the floor and run into the drain.

Preparing the patio

For basement floors, you used the walls for the framing of the concrete. For the patio, you need to build your own frames or molds.

When all the digging is done, outline the patio area with stakes that are the exact height you want the patio to be. Use a stake at all four corners and every 8 feet along the perimeter of the patio. Use a line level to be sure that all the stakes are the proper height.

Level lines   Level lines

Use 1-x-5 or 1-x-4 boards to make the framing. Position each board so that the top of the board is level and aligned with the tops of the stakes. Check all the boards to see that you have a level horizontal reading all around the patio.

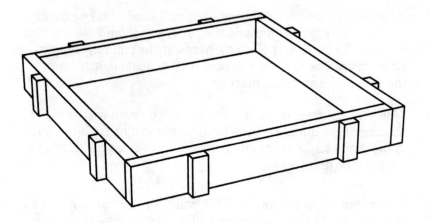

If you plan to let the cement-block basement wall of the house serve as one boundary for the patio, strike chalk lines along the blocks to give you a pouring line. Do not let the line rise higher than the doorway at any point.

Fill the bottom of the patio area with sand and gravel, then wait for the concrete. You do not need to use plastic here because the patio is outdoors and moisture does not matter.

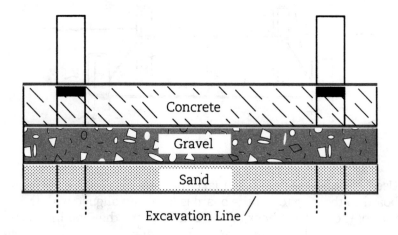

Concrete

Gravel

Sand

Excavation Line

If you choose to do all the concrete work yourself, you must rent a mixer or be prepared to devote many hours of hard work to mixing concrete. Have sand delivered or shovel it yourself from a creekbed or other natural source. It does not matter if the sand has pebbles in it because the concrete needs a *coarse aggregate* (gravel and similar materials) as well as a *fine aggregate* (sand).

When you mix concrete, a ratio of 1:3:5 is commonly used. This means for one part of cement mix use three parts of sand and five parts of gravel or other coarse aggregate.

Mix in a manner similar to that for mixing mortar. Start with 5 gallons of water, add seven shovels of sand, then add half a bag of cement mix, and when this is thoroughly mixed, add the rest of the sand.

Mix completely. Add gravel only when the other ingredients are integrated into a plastic mixture. If your sand is too wet, you can add slightly more sand and gravel and a shovelful of cement mix.

Keep a bucket of water handy. When you start to shovel the mixed concrete into the wheelbarrow, it is a good idea to wet the wheelbarrow thoroughly before you shovel in the concrete. You should wet shovels, hoes, rakes, and trowels before using them in either concrete or mortar.

Haul the wheelbarrow load of concrete to the patio or basement area. You should mix the concrete as close to the pouring site as you can and try to work near a ready supply of water.

Your *forms* (the framing for the concrete) should be ready if you are using them. You may find that it is easier to section off a small part of the basement before you start to pour.

If you have a small room, start here. Do not learn to pour concrete in a broad expanse of floor where mistakes you make will be evident to anyone entering the room. Start in the small room and begin dumping the concrete in the far corner. As soon as you dump it, use a trowel or smoother to start shaping the concrete to conform with the line on the wall and with the stakes.

Pouring your own concrete

As you work your way toward the doorway, pour until you can barely reach the wall. When you are at this point, you can start to apply the finishing touches to the concrete.

Screeding and floating the concrete

When the space is ready, begin the *screeding* process. This is the initial step in finishing concrete, and your first measure is to use a strike-off board. To strike-off concrete, two people are needed to hold and maneuver the board, which is stood on edge and worked in a sawing motion across the tops of the form boards. In the basement there are no form boards, so you use the strike-off board to work the concrete down to the point where the chalked line and the tops of stakes are visible.

As you work, keep a sufficient amount of concrete in front of the screed board. For small areas, the board can be a 3-foot length of 2-x-4 or similar timber.

The concrete in front of the strike-off board guarantees you that the minimum depth is being achieved by the concrete and that you are raking off surplus only after the necessary concrete has been maneuvered into the form area. As you near the door, add a board across the doorway as a form.

When the screeding process is complete, you are ready to *float* the concrete. You can use a metal or wood float to get the smoothest surface. Use a rectangle of plywood (2 feet by 4 feet or even 4 feet by 4 feet) and kneel on the plywood while you work. You may need to lay wide boards across the concrete for you to reach the back corners without wading in the concrete.

While you kneel on the plywood, reach forward and use the float to rub the top of the concrete gently, pushing the float back and forth as the gravel is worked slightly deeper and the moisture is forced to the top. Lower high places, fill in low places, and bring enough mortar to the surface to keep a smooth finish. You can use a level or straight-edged board or timber (a 4-foot length of 2-x-4 works well) to lay across the concrete. If you can see any spaces under the bottom edge, use the float to correct the problems.

You can buy a long-handled float (sometimes called a *bull float*) that will permit you to float the surface of the concrete without having to walk on it or use the plywood support.

As you complete a section of the room, move your support plywood back and continue operating. Do not stop until you reach the doorway where the modified form keeps the concrete from settling after your work is done. If you plan to stop at the doorway, install the form board halfway across the doorway opening. Later you can install the threshold to cover the mark where the operation stopped.

For larger areas that cannot be done in one day, section off a part of the room by using form boards. Pour and finish as before.

You should take care that the concrete does not dry out too rapidly. Concrete begins to set up in one hour. After a full day, it is hardening to the point that it will not give when you step or push down on it. Complete curing takes about seven days. During the curing time, the concrete should be kept as moist as possible. You can spray water on it several times a day or cover it with wet burlap or similar materials. Fresh concrete should be protected from direct sunlight for a week if you can arrange proper covering.

As you continue to work, try to stop each day at the least noticeable places. If you know where a partition wall will be added later, install form boards along the wall line and plan to stop at that point.

Your pouring of the basement might take a week or longer. You can save well over $1000 during this time, so unless you could earn $1000 during that week, you are saving considerable money by pouring the concrete yourself.

There is a great deal more to pouring concrete than the simple steps outlined here. These procedures will work for your patio and basement.

As concrete dries, it actually hydrates the cement in the mixture. Technically, it does not dry. Engineers have said that concrete continues to harden for years after it is poured. Your basement-floor surface will scratch easily for weeks after you finish it, so be careful about dragging heavy objects across the floor soon after the concrete is hard enough to walk and work on.

Your next step is hanging drywall or installing wall covering.

Installing drywall ceilings

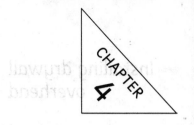

WHEN YOU ARE ready to install ceilings, you have a choice of drywall (sometimes called Sheetrock plasterboard or gypsum board), tongue-and-groove boards, ceiling tiles, and several other materials. This chapter is concerned with drywall ceilings.

Make a quick visual check of the ceiling joists before you begin work. Hold a straight edge against the bottom edges of the joists to see if any of the joists have bowed or sagged. The bottom of all joists should touch the straight edge.

Ceiling joists

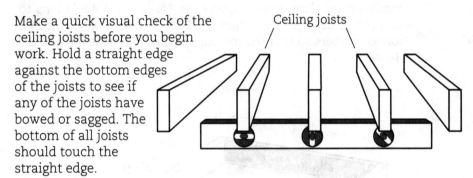

If all is correct, you are ready to work. If you have not bought materials, you can get a good estimate of the number of panels you need by multiplying the dimensions of the room and dividing by 32, which is the number of square feet in an 8-foot panel of drywall. If the room is 20 feet by 24 feet, you have a total square footage of 480. Divided by 32, this means you need 15 panels of drywall.

Waste occurs in almost every type of construction. Most carpenters or builders add 10 percent to the order to allow for waste.

Start in any corner and use a square to see if the corner has retained its original trueness. If the corner is perfectly square, you are ready to install the first panel.

Estimating needed materials

Installing drywall overhead

If you are working with helpers, the simple way to install drywall overhead is to have two people, one on each end, to lift the panel and hold it in position while you nail or screw it in place. The three of you may need to stand on short stepladders or low scaffolding while you work.

If you are working alone, you may need to construct a stiff-knee device that holds the panel in place while you work. You can make a simple supporting helper from two long 2-x-4s and four shorter 2-x-4s. Measure the distance from the bottom of the ceiling joists to the subflooring. Then deduct the thickness of the drywall plus ¼ inch to ½ inch of free play. Lay the two long 2-x-4s (as long as the measurement you just made) on the floor, standing on edge. Nail a 2-x-4 4 feet long to one end of the 2-x-4s. Position the 4-foot 2-x-4 between the two longer timbers so that the top edge is flush with the ends of the 8-foot 2-x-4s.

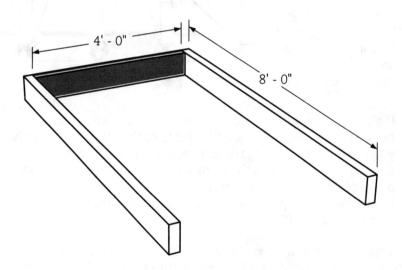

Do the same at the other end of the 8-foot 2-x-4s. Now nail the other two 4-foot 2-x-4s to the outside edge of the longer timbers and parallel to the wide side of the 2-x-4s.

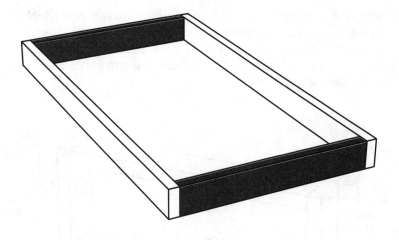

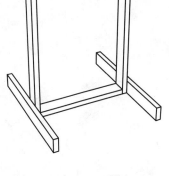

You now have a structure that stands erect and reaches nearly to the bottom of the ceiling joists. To make your work much easier, knock together a second structure like the first one.

Using the stiff-knee

When you are ready to use the device, stand the first support or stiff-knee in one corner and as close to the wall as you can get it. Stand the other against the same wall so that the supports are close to each other.

Carry a panel of drywall close enough to it so that you can lean the top end of the panel against the support device. Lift the panel and push the end over the top edge of the corner 2-×-4 crosspiece at the top. Work the panel into position so that you can slip the other end on top of the second support.

The panel is now ready to be nailed. You have a small amount of play between the panel and the bottom of the ceiling joists, and if you need to, you can wedge a thin shim of drywall or smooth wood under the panel on each of the support pieces.

Constructing a framed helper

If you do not want to be troubled with shims or the play between panel and joists, you can devise another helper that will eliminate all problems. You need four lengths of 2-×-4 5-feet long. You can also use 8-foot 2-×-4s if you don't have scraps available.

Locate a 4-foot length of 2-x-4 and nail it between two 1-foot lengths of 2-x-4 so you have an inverted and squared U. Nail the 8-foot 2-x-4s to four lengths of 2-x-4 4 feet long that run parallel to the wide side of the timbers and can serve as feet for the device.

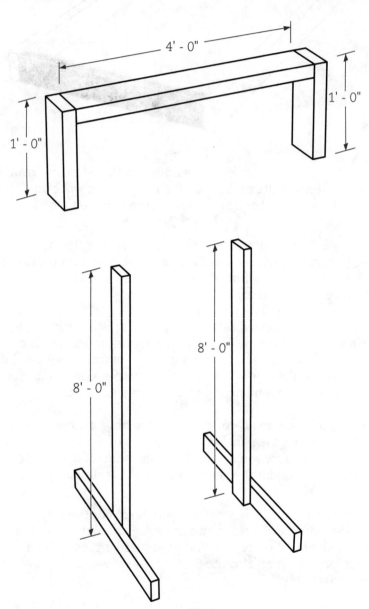

Tack strips of wood across the 8-foot 2-x-4s so that they stand together. You can use a strip at the bottom and one near the top.

You need four medium-size C-clamps. You are not wasting money when you buy these. They are inexpensive and have an astonishing number of uses. Look at it this way. You can buy all four C-clamps for what it would cost you to pay a helper to work two hours.

Place the assemblies close enough so that the drywall panel can reach from one to the other. Position the first of the inverted U structures so that the downward-pointing 2-x-4s are against the longer timbers. Use a C-clamp on each side to clamp the top assembly to the bottom one.

Do the same at each end of the drywall holders. Now lift a panel and slide it on top of each of the holders. If there is play, loosen the C-clamps, one at a time, and push upward until the panel is held tightly against the ceiling joists. Reposition all four locations if necessary. You can now nail or screw the panel into its permanent position.

It may seem like a lot of trouble to construct these devices. In actuality it takes only a few minutes to do the work, and they can save you hours of work and frustration while you are drywalling the ceilings of your house.

If you use screws to fasten the drywall, you can start the screws with a traditional screwdriver, then finish them with a power screwdriver tip in your electric drill.

Installing drywall with screws

Attaching the drywall with screws takes more time, but in the long run you benefit. The screws will not work loose from the wood and the drywall will remain securely attached to the ceiling joists.

Be sure to install the drywall so that it crosses the joists the longest way. Each panel should cross seven joists.

To install the second panel, lift it to the supporters and position it exactly before you raise it to its permanent place by adjusting the C-clamps. Butt the end of the second panel against the end of the first panel. Nail or screw the piece into place.

If you use nails, space them no more than 6 inches apart as you move along the joists. Be sure that the panel is nailed to every joist. If you use screws, space them 12 inches apart.

Sinking nails and screws properly

Drive nails or screws until there is a slight indentation or dimple in the drywall surface. Do not drive them deeply enough to break the paper. Drywall is not strong in tensile strength, and it is weakened considerably if the paper is damaged or torn badly.

Complete the first course of panels. Install the second course exactly as you did the first. Continue across the room until you reach light-fixture locations. You must cut out for these boxes, and unless you work carefully, you can ruin panels of drywall or wind up with an unattractive, ill-fitting panel.

Installing around wall irregularities

The easy way to install the panels over the fixture boxes is to lift a panel onto the holders, position it carefully, then push up in the vicinity of the fixture box. The pressure will cause an imprint on the back of the panel so that you can cut out the imprint and have a good fit.

You can also measure from the last panel installed to the near side of the fixture box, then measure from the edge of the near panel in the previous course. Turn a new panel facedown and measure and mark for the location.

Take a fixture box and place it on the panel, locating it so that the sides touch the marks you made. Hold the box in place and outline it with a pencil. You can now cut out the outline of the box. Use a pocketknife to make a neat slit along one line. Slip a hacksaw blade or a keyhole saw into the slit and cut out the outline. When you lift the panel into place, you will have a perfect fit if you measured and marked correctly.

A third method is to use a section of waste panel with a factory edge or a very straight-cut edge. Hold the section so that it butts into the last panel installed and one end of the section covers the box. Apply pressure as before and cut out for the outline. Then lay the section on top of a new panel and make the pencil outline. Cut out the outline as you did before and install the panel.

Continue working across the room until you reach the other side. You may not have room for a full panel. If you need to cut panels, measure from the wall to the edge of the near panels that have been installed. Mark and cut the new panels to fit, less ¼ inch.

It is always a temptation to try to get the closest fit you can. In many instances this is a mistake, particularly if the fit will not be visible when the finish work is done. If the fit is too tight, you need to hammer or wedge the piece into the space, and you can damage one or both of the pieces. It is better to have a slight amount of working room against walls or where molding will cover the joints or seams.

When you cut a panel, turn it facedown if you use a handsaw. If you use a razor knife, lay the panel faceup and mark the cut line with chalk or pencil and straightedge.

Cut into the panel with the razor knife so that the blade penetrates the paper fully. Draw the knife slowly and carefully along the line to the end of the panel. Move the panel so that the part to be cut away hangs over the edge of the work surface. A good idea is to use two 2-x-4s under the panel and align the cut with the outside edge of one of the 2-x-4s.

When the cut has been made, hold the cutaway section firmly in one hand and push down while the other hand holds the usable portion of the panel in place. The panel will break along the cut line, and you can use a knife or razor knife (or scissors if you wish) to cut the backing paper.

Getting a proper fit

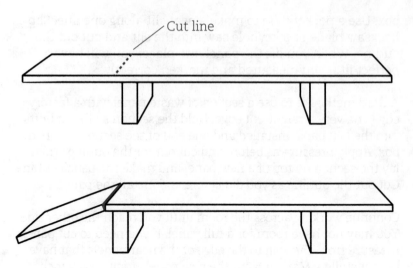

When the room is complete, you are ready to apply the compound and tape. See chapter 6 for a complete discussion on the use of compound and tape and for information on sanding finished compound work.

Installing tongue-and-groove ceilings

CHAPTER 5

DRYWALL ceilings look fine in many houses and serve their purpose well. Some people do not have the stamina or strength to put up such ceilings and need material that is less difficult to handle.

Start by choosing your lumber carefully. This material is not cheap; in fact it is rather costly, but in the long run it lasts for years and years and retains its original charm and beauty.

When you select the lumber, inform the dealer that you need to help select the merchandise. He or she may object, but you should remain adamant and insist that either you will pick out the lumber or you will return any units that are defective. If the dealer does not agree, look for one who is more understanding. Tongue-and-groove lumber is very beautiful. It is expensive, and it cannot be used effectively and economically for anything but ceiling and wall coverings or flooring. If the tongue or the groove is badly damaged, you cannot use the piece of lumber, and your money is wasted. Do not buy a product that cannot be used.

Your choices are varied. You can get knotty pine and a number of similar types of ceiling and wall coverings. You also have a choice of standard tongue and groove or V-joint tongue and groove. You can buy it in several widths, but the 1-x-4 or 1-x-5 works best.

Choosing materials

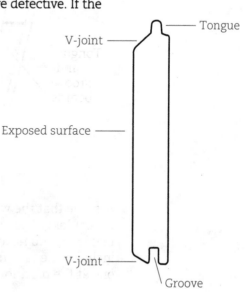

Tongue

V-joint

Exposed surface

V-joint

Groove

Installing tongue-and-groove ceilings 43

A knotty-white-pine V-joint tongue and groove makes one of the most attractive ceilings. You can calculate how much you need by remembering that if you are using 1-x-4 lumber, you need 3 linear feet to equal 1 square foot.

Estimating materials

Multiply the length of the room times the width to find how many square feet you need to cover. A 20-x-24-foot room has 480 square feet in it. You need 1440 linear feet to cover the ceiling, plus 10 percent waste allowance, which results in 144 additional feet. You need 1584 linear feet, but you can order 1600 linear feet to be sure that you have enough lumber to cover the ceiling.

Ask for the lumber in a variety of lengths, keeping the size of your ceiling in mind when you make the choice. Remember that you do not want to let ends of units occur on the same joist or rafter in succession.

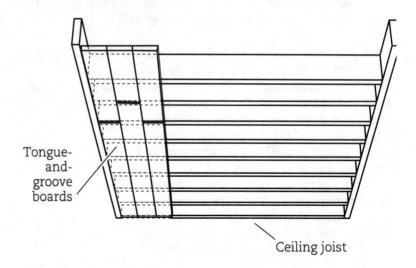

Tongue-
and-
groove
boards

Ceiling joist

Assume that the wall that runs parallel to the joists or rafters is 20 feet long. The wall at right angles to the joists or rafters is 24 feet long. You need to cross or nail to a total of 18 rafters. This includes the rafters at the beginning of the room and the final one at the partition wall. A 12-foot board will cover half of one course in the room. You need two 12-foot boards to reach all the way across the room. You can also use three 8-foot boards.

Your order becomes simple in this light. Order half 8-foot boards and half 12-foot boards. A 20-foot-wide room has 240 inches across. The 12-foot boards will cover 10 feet. It will take six of them (if you use 1-x-4 boards) to cover 1 foot of the wall. You need to cover 10 feet of the wall, so you need to order 60 of the 12-foot boards. It will take 9 of the 8-foot boards to cover a foot of linear ceiling space, so to cover 10 feet, you order 90 8-foot boards.

When the lumber arrives, see that it is unloaded with care. Do not stack it so that it is in a bind or strain. Lay several 2-x-4s across the floor of a protected room and stack the lumber across the timbers. There should be at least three 2-x-4s under the lumber.

Caring for lumber

Stack the 12-foot boards in one area and the 8-foot boards in another. You need to get to the lumber readily, and you do not want to shuffle through the stack in search of a particular length.

When you are ready to work, begin with a 12-foot board pushed firmly against the corner and against the 24-foot wall. You have noted by now that when you ordered the specific lengths, you virtually eliminated your waste. You did not only eliminate waste; you saved a great deal of time. You will not need to stop nailing to measure, mark, and saw lumber.

Use *cut nails* to install the boards. You also need a supply of finishing nails 3 inches long.

If you have a helper, have him or her hold the board in place while you nail it. It is very difficult for one person to hold a 12-foot board in position with one hand and drive nails with the other.

Installing ceiling

If you are working alone, locate a 2-foot length of scrap (a 2-x-4 or 1-x-4 works well) and a small block of 2-x-4 about 6 inches long. Nail the 1-x-4 or 2-x-4 to the 6-inch block so that the ends are flush. Now go to a point midway along the rafter span for the 12-foot board and nail the block assembly to the bottom of the joist or rafter so that the 2-foot board is separated from the joist by the 6-inch block.

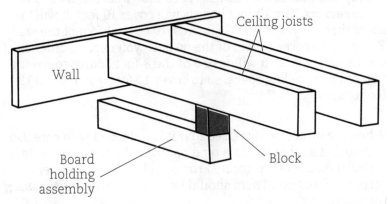

The assembly should be positioned so that the open end of the assembly touches the wall where you are planning to start work. Push one end of the ceiling board through the space between the joist and the 2-foot board. The 2-foot section holds the end of the ceiling board in place while you nail the other end.

Start with the board positioned so that the groove is against the wall side of the room and the tongue is facing you as you work. With the board pushed firmly against the wall framing, drive a finishing nail straight up into the board and into the joist. Drive the nail an inch from the wall framing.

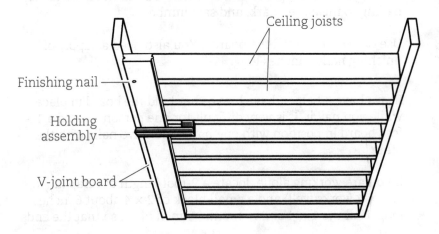

Do this at every joist, including the first and the last the board spans. The back side of the board is now secured firmly to the ceiling. You can pull down the holding assembly and complete installing the board.

On the front side of the board (the side near you as you work) drive a cut nail at an angle through the ceiling board and into the joist. Start the squared point of the nail at the angle formed by the tongue and the edge of the board.

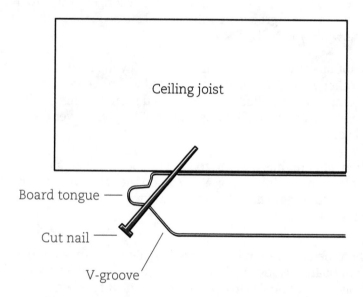

Slant the cut nail so that the nail is pointed toward two o'clock. Drive the nail in so that the head is even with the outside edge of the tongue. Do not attempt to drive it any deeper at this point. If you damage the tongue, you will have great difficulty in trying to fit the groove of the next board on the tongue.

Keep a *punch* handy. When the nail is driven even with the tongue, use the punch to drive it the rest of the way. Set the punch so that the point is centered upon the head of the nail and strike the end of the punch until the nail head is fully seated and even with the edge of the board. You can also turn the punch sideways and lay it across the nail head. Hold the

Sinking nails properly

punch by the scored portion of the shank and strike the side of the punch to sink the nail.

When the first board is completely installed, its end should reach halfway across the joist edge where the board stops. The next board is butted against the end of the first board and installed the same way.

Using a board holder

You need to use your board holder to start the second board. You do not need the holder assembly any longer in this room. The tongue-and-groove fit will hold all subsequent boards in place while you nail.

Do not use any more finishing nails until you are all the way to the other wall. The first ones you used will be covered by molding to be installed later. You need more finishing nails for the very last course of boards, and molding will cover these nails as well.

When you are finished, the room will not have a single nail visible in the entire ceiling. You can do the same with the walls as well, so that no nails are visible anywhere except in molding and window facing.

Installing the second course

When the two 12-foot boards are installed, nail up a course of 8-foot boards. Notice that the 12-foot boards ended on a rafter or joist so that no sawing was needed. The 8-foot boards also end on a rafter, and again you do not need to saw the board.

It is for convenience like this that this book stresses the need to plan carefully. Do not think only in terms of space or layout of the rooms. Think of the time and money saved. If the room had been 13, 14, 15, 17, 18, or 19 feet long, you would have needed to saw boards or to order extremely varied lengths, which are usually unavailable. Only 12-, 16-, and 20-foot walls or ceilings (unless you have larger rooms) let boards end on rafters.

If you had ordered 10-foot boards, you would have at least 1 foot of waste on each board. You would also have to take time to measure, mark, and cut—all of which steals building time if not necessary parts of construction.

Nail up the 8-foot boards when the 12-foot course is complete. Lift the board and position the groove over the tongue. Push gently, raising and lowering the other end of the board as needed to get the desired fit.

Correcting poor fits

If a board is obviously badly warped or curved, lay it aside. If you cannot get the tongue and groove to fit perfectly, take the board down and try another.

When you leave a poor fit, you will have trouble with nearly every board after that point. Keep the fit flawless, and your work will be faster, easier, and neater.

If a board is apparently straight but the tongue and groove does not fit no matter how you try, examine the groove to see if it has been damaged. Sometimes the bottom edge becomes splintered or damaged. Examine the tongue as well to be sure that it hasn't been smashed at some point or badly damaged.

Preventing damage to tongues and grooves

If a splinter on the back side of the board will not allow the fit, break the splinter off, and the spot will not show when the board is installed. If the groove is damaged, you can often cure the problem by using a small length of tongue and groove lumber to run up and down the blemished spot. Place the short scrap so that the tongue is inserted into the groove and push the scrap length back and forth several times to clean out the groove. If the tongue is damaged, turn the scrap over and insert the groove over the tongue. Repeat the process. Sometimes this straightens a damaged tongue.

If all else fails, use your pocketknife and simply carve or cut out the problem area, which is often less than an inch long. Cut down completely through the tongue so that the problem is totally eliminated. This correction will not show if you do not cut into the board itself.

When the two boards fit almost perfectly but you have trouble seating the new board the rest of the way, place a length of scrap board on the tongue of the new board at the problem area. Strike the scrap board with a hammer and drive the new board the rest of the way into position.

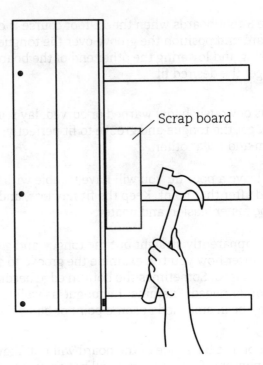

Scrap board

As you move across the room, you might encounter boards that are slightly curved, perhaps not enough to render them useless but enough to keep them from fitting. Sometimes these boards are 3 or 4 inches from true at the end of the board where it is supposed to fit into the installed board.

You can often seat such boards by using modest leverage. Go to the beginning of the problem area and 4 or 5 inches from the outside edge of the board use the C-clamp to hold a length of scrap 2-x-4 tightly against the joist or rafter. Insert a short length of scrap lumber so that the groove fits over the tongue. Then, with the outside edge of the scrap piece an inch or so from the end of the 2-x-4 held by the C-clamp, insert the tip of a pry bar and apply pressure.

If you see that the plan will work, as it usually does, start the nail into the tongue area and apply pressure again. When the fit is perfect, hold the pressure on the board and drive the nail in completely. Move the C-clamp toward the end and repeat this measure as often as needed. At times it works better to apply pressure at the end of the board first. If one method does not work, try another.

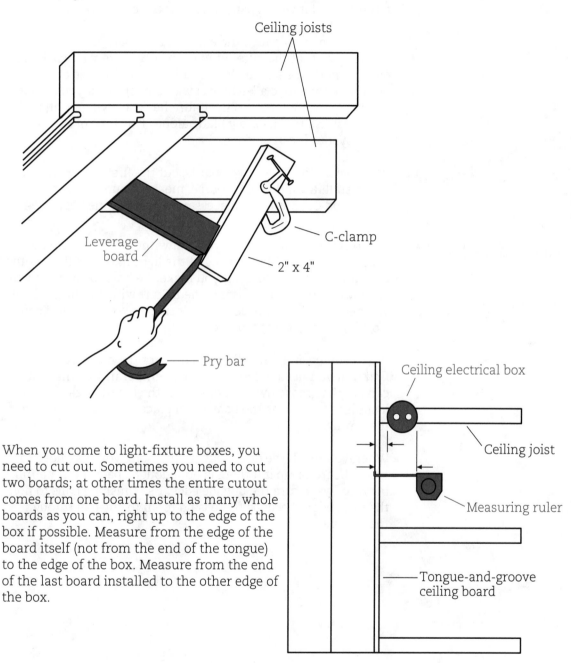

Ceiling joists

Leverage board

C-clamp

2" x 4"

Pry bar

Ceiling electrical box

Ceiling joist

Measuring ruler

Tongue-and-groove ceiling board

When you come to light-fixture boxes, you need to cut out. Sometimes you need to cut two boards; at other times the entire cutout comes from one board. Install as many whole boards as you can, right up to the edge of the box if possible. Measure from the edge of the board itself (not from the end of the tongue) to the edge of the box. Measure from the end of the last board installed to the other edge of the box.

Mark the new board according to the marks, then lay a new electrical box in position on the back side of the board and trace around it with a pencil. Cut out the outline.

If the outline is between the edges of the board, drill a hole carefully so that the outer limit of the drill barely touches the outline mark. Insert the blade of a jigsaw in the hole and cut out the outline. Use a keyhole saw or a hacksaw if you do not have a jigsaw. Sometimes the entire hacksaw will not fit, and you must use the hacksaw blade alone. This type of sawing is not easy, but it can be done.

When you need to cut two boards, cut the first one as needed and install it. On the next course, measure and mark as you did before and cut out the unnecessary portion of the board.

Getting perfect fits

Try to get the fit as close as you can without being extreme about the matter. The housing of the light fixture will cover the cutout if it isn't too loose. If you make the hole too large, it will show when the housing is installed. Start with a tight fit and if you need to, shave a little off the inside edge of the cut. Keep working until you get a proper fit.

When you begin to install the board over the box area, you may find that you need to loosen the box temporarily so that you can get the groove over the tongue of the board already installed. As soon as the board is in place and nailed, tighten the box again.

When you reach the other side of the room, you may need to trim a board or rip it if there is not enough space left for a full board. Always trim the outside edge. Use finishing nails to hold the final board because there is no room to cut-nail through the tongue.

Drywall installation

AFTER THE CEILING covering has been installed, you are ready to start on the wall coverings. This is easier than ceiling installation and moves much faster.

You have a variety of wall coverings from which to choose. Among the most popular are drywall (gypsum board or Sheetrock plasterboard), paneling, and tongue-and-groove boards.

If economy is your primary goal, drywall may be the best choice. This choice also provides you with one of the fastest installations. Drywall comes in 8-foot and 12-foot lengths and 4-foot widths. For most houses, the 8-foot height and 4-foot width works very handily. You can buy drywall from ¼ to ⅝ inch thick. You can also buy it with beveled, tapered, or square edges.

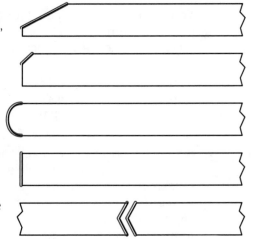

Each panel or sheet of this material covers 32 square feet. To figure out how much you will need, determine the number of square feet in the room and divide by 32.

If a room is 20 by 24 feet and it has three windows and one door, you calculate your drywall needs in the following manner. The ceiling is 8 feet high, so you will have two walls that are 20 feet by 8 feet, which gives you 160 square feet for each wall. Twice that amount is 320 square feet.

Estimating necessary materials

The other two walls are 24 feet by 8 feet, or 192 square feet per wall. This equals 384 square feet for the two walls. The total square footage of the wall area for the entire room is 704. Divide 704 by 32 (the space covered by one panel) and you get 22, the number of panels you will need.

You can deduct the square footage of the door and windows if you wish.

If the door is 3 feet wide and 6½ feet high, you have 19.5 square feet to be deducted. The three windows, we assume, have 15 square feet each, so the three windows will equal 45 square feet. Add the 45 and 19.5 square feet and you get 64.5 square feet to be deducted from the total, or roughly 640 square feet. This is equal to 20 panels of drywall.

You can buy an extra panel for the room if you think there is a danger of damaging one, or you can return to the supplier and buy extra panels if you need them.

When you are ready to install the drywall, you need a hammer, a razor knife, and some *drywall nails*. These nails have broad heads and are shanked to keep them from pulling out under stress.

Checking studding construction

Before you begin work, check the studding to be sure that all are straight and square. Hold a tape measure in one corner and quickly run it over the studding. As you know by this time, 16-inch on-center spacing is standard, so your studs should fall on 16-inch intervals. Many measuring tapes or rules have stud locations clearly marked. Some are labeled *Stud*; others have the numbers in red or other highlighted colors.

Check at the top, middle, and bottom of each stud. If a stud has warped or bowed badly, you may need to replace it. If there is a slight bowing or bending, you can use a *spacer board* to correct the problem.

Where the stud bows excessively, cut a length of 2-x-4 that is 14½ inches long. Insert it into the space between studs with one end against the next stud and the other end against the bow in the defective stud. Tap the unit into place. As it moves closer to horizontal, the stud will begin to straighten. When a correct position is achieved, nail through the studs and into the end of the spacer.

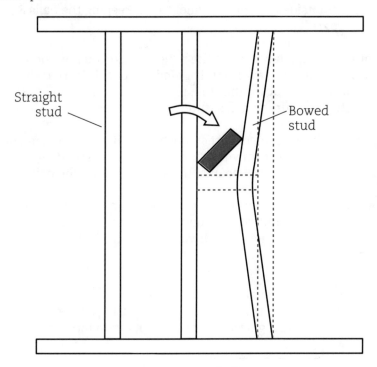

Straight stud

Bowed stud

If the stud bows toward the center of the room, you cannot correct the problem. Remove the stud and install another in its place. When all studs are correct, you are ready to start hanging the drywall.

Beginning installation

Start hanging in a corner. Your most important consideration is to start with a square corner, which should be no problem if you framed the house correctly. To be certain, stand the first panel in the corner and move it so that it aligns with the corner framing as closely as possible.

If you have a correct fit, install the panel by driving nails through the drywall and into the stud behind it. The first nails can be placed at the top so the entire panel remains rigidly in place while you nail the rest.

Nail through the panel and into the top plate or top cap. Keep the nails at least an inch from the edge of the panel. Drive nails no more than 8 inches apart across the top, then drive nails down the outside edge of the panel, again keeping the nails 8 inches apart.

Continue completely around the panel. When you have done this, drive nails into every stud behind the panel. Start at the top and continue down the stud, placing nails 1 foot apart. At the bottom, be sure to nail securely to the sole plate.

As you work, start the nail, then push in with your free hand on the panel until it is firmly positioned against the studding. Do not try to pull the panel against the studs by using the nail.

You can also buy screws to install the drywall. The progress made by using a traditional screwdriver is very slow, so you may need to buy some screwdriver heads that fit the chuck of your power drill.

Using adhesive with drywall

A third method of installing drywall is to use a special glue or adhesive. You fit the panels well first, then, using a type of caulk gun, apply a bead of glue along the studding from top to bottom.

When the beads are applied, lift the panel and set it into position. Push it firmly against the glue or adhesive and drive a nail along the top and bottom to keep the panel in constant contact with the studding.

If you don't want to bother with filling nail holes later, you can construct a simple bracing device that will hold the panels in place briefly. Use a section of plywood or wallboard 4 feet wide and 3 feet high to make the brace. You will also need two lengths of 2-x-4 or similar bracing timber. Attach the plywood to the ends of the 2-x-4s and nail some blocks of wood to the subflooring. Push the brace panel against the drywall and hold it there while you nail the other ends of the brace to the wood blocks.

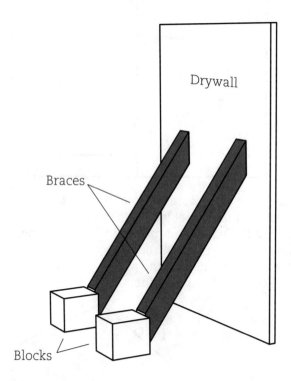

The adhesive dries rapidly. Leave the bracing in place for only a few minutes, and then take it down.

Adjusting corners for proper fits

If you do not have a good fit, push the panel into the corner until one corner strikes the framing. Put a level on the outside edge of the panel and see if you have a true vertical reading.

Where the ill-fitting edge strikes the studding or corner post, mark with a pencil on the sole plate or the top plate. Take the panel down and measure from the mark to the corner.

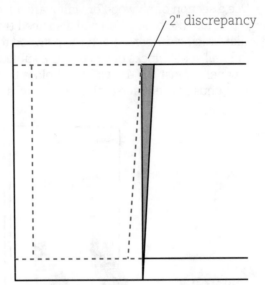

2" discrepancy

Lay the panel on a work surface and mark it as follows. Assume the top corner of the panel lacked 2 inches to fit exactly. Go to the bottom of the panel and mark 2 inches from the edge. Use a chalk line and strike a mark from the corner position (2 inches from the edge) diagonally to the point of the upper corner. Use a razor knife to cut along the line.

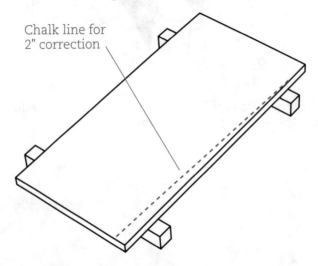

Chalk line for 2" correction

For best results, you may want to hold a straightedge, like the blade of the square, on the chalk line while you cut along the line. Be sure to hold the blade of the square rigidly.

With a little practice, you can learn to cut along the line without the use of any straightedges. The point of the blade bites deeply into the panel and keeps you from letting the blade veer too sharply.

If the cut is not completely through the backing, grasp the part to be removed and with a twisting motion snap it so that only the backing is holding. Use the razor knife to cut through the backing, and the piece will fall free. You can use a circular saw or handsaw to cut drywall, but the accompanying dust is stifling. The razor knife works best.

Stand the panel back in place. The fit should be perfect now. Nail the panel in place.

You now have a square start. The remainder of the panels should fit well and with a square fit. To install the remainder of the panels, simply butt the edge of one panel into the edge of the adjacent one and nail the panel in place. Each panel should end halfway across the edge of the stud.

When you reach the next corner, if a full panel will not fit into the space, measure from the edge of the last panel installed to the corner. Cut the amount from a full panel and install, letting the cut edge fit into the corner.

When you start the next wall, check again to see that you are off to a square start. If not, fit the panel as you did earlier, then proceed with the rest of the wall as described.

Installing around windows and doors

When you come to windows, you may find that a panel ends against the window framing. This is fortunate because it makes the rest of the work easier. If it doesn't work out that way, measure from the end of the nearest installed panel to a point 16 inches from the outside edge.

Next measure from the ceiling joist down to the top of the window framing. If this distance is 14 inches, mark the panel accordingly. Measure up from the floor to the bottom of the

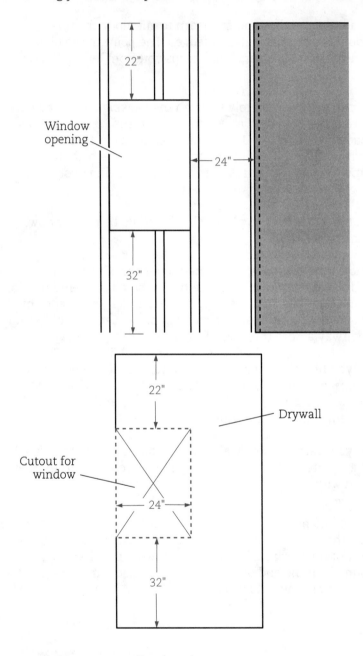

window frame and mark the distance on the panel. Connect the marks with a chalk line or straightedge. You will not have a long rectangle to be cut out of the drywall.

Lay the panel on the worktable and cut along the marked lines with a razor knife. If you need to, complete the cut with a pocketknife or a handsaw.

Remove the cut-out section and stand the panel in place. It should fit into the remaining space neatly. Nail it in place and repeat the same procedure on the other side of the window.

Use this method of installation over doorways. If you must cut a narrow strip to fit behind a door, fit the factory edge against the last panel installed and let the cut edge fit into the corner.

Compounding and taping joints and indentations

When you have completely covered the walls of the room, you are ready to begin compounding and taping. When you drive a nail, hit the head of the nail one final time after the head is flush with the drywall surface. This last blow makes a small indentation in the surface. You need this indentation. If the nails are flush with the surface, you will have difficulty applying the compound.

Use a putty knife to apply the compound. Dip the tip of the blade of the putty knife into the compound and scoop up a small amount of the mixture. Turn the putty knife so that you can rake the compound over the indentation at each nail. Smooth the compound so that the nail head is completely covered and the indentation filled, so that the surface of the panel is perfectly smooth.

Where two panels join, you will notice a slightly recessed and slanted area near the edge. Use the putty knife to fill these places. Then cut a length of drywall tape and, starting at the top, embed the end of tape into the compound. Move down the seam slowly, pressing the tape into the compound firmly the entire length of the seam. Then go back and cover the tape with compound.

As you complete a seam, "feather" the compound by spreading it thinner and thinner until there is no compound left under the putty knife. Do this with every seam and every nail hole that has not been filled to the level of the panel.

Finishing the surface

Let the compound dry overnight. The next day, use sandpaper to sand the filled areas until they are perfectly smooth. You should be able to run your hand over the seams and detect no difference in the surface levels. The same is true of the nail-head areas.

It is important to wear a dust mask and protective eye covering while you work. Dust from the compound can be harmful to your lungs and eyes. Some people are allergic to the dust if it touches their skin, so you may need to wear long-sleeve clothing.

If you have outside corners in the room, you can buy special corner bead material that fits over the corners and is stronger and more durable than tape. When the bead is nailed into place, use the compound and putty knife and compound it as you would tape.

If you wish to improve the room further, you can use a single thickness of wallboard or drywall for your initial installation and add a second layer. This second layer offers great insulation and soundproofing qualities, strengthens the wall, and adds greatly to the fire-control properties of the room.

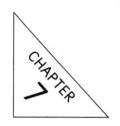

CHAPTER 7

Tongue-and-groove wall covering

YOU HAVE ALREADY seen that you can cover ceilings with tongue-and-groove boards at a fairly economical price and with speed and ease. You can cover walls in the same manner.

A very attractive look in a modern house is ceiling and walls of the same knotty-pine, V-joint, tongue-and-groove lumber. You will have molding, window facing, and flooring to break the tendency toward sameness that you might not want to push too far.

Estimate your lumber needs as you did with ceiling boards. This time you will need more lumber, and you will need to think of more waste. Because of the dimensions of the room, there was essentially no waste in the ceiling installation. You may find more waste unavoidable in wall covering.

Return to the example we used for wall covering. The room is 24-x-20, with a door to the hallway, one window on each of the outside walls, and a door to the bathroom. There is also a huge closet with double-wide doors on one wall.

Estimating lumber needs

Your first consideration is how much lumber to buy. Keep in mind that if you buy huge amounts, there is a greater possibility of damage, theft, and vandalism. You might also spend money before there is a need to do so. By buying a little at time, you can pay for the materials and labor expenses as you work. In this fashion, you will have the house essentially paid for when it is complete.

The walls in the room are 8 feet high. This means that you will need three courses of 1-x-4 boards to cover 1 linear foot of wall space.

You have 480 square feet in the room. (If your rooms are not the same dimensions you can simply substitute your dimensions and solve the problem in the same manner.) The first exterior wall has 192 square feet in it, as does the opposite wall. That is a total of 384 square feet. The other two walls have 160 square feet each, for a total of 320 square feet.

The four walls have a combined 704 square feet of wall space, less the space occupied by windows and doors. Tell the dealer of the dimensions of the room and the 704 square feet, less wall interruptions. He or she can usually tell you how much lumber you need.

For the first exterior wall, you have a 24-foot length and an 8-foot height. You will cover 4 feet of height with 12-foot boards and 4 feet of height with 8-foot boards. You need 2 12-foot boards per course and three courses of boards per foot, which means you must have 6 boards per square foot and 24 boards for the 4-foot height of the room area.

You need 3 8-foot boards per course and three courses per linear foot, which means 9 boards per linear foot. This in turn amounts to 36 boards for the 4-foot section of the wall that will be covered by these boards.

Figuring walls with openings

The window in the room is 60 inches high and 42 inches wide (again, substitute alternate figures for your own needs). This works out nominally to 5 feet by 3½ feet, or 17½ square feet that you can deduct from your lumber order.

Your best plan is to order enough lumber to board the entire wall. The window area will probably be taken up by the waste. You had little if any waste in the ceiling, but the walls will be different.

The bathroom rough door opening is 78 inches high and 26 inches wide, or 16.9 square feet. Closet doors are 4 feet wide and 80 inches high, which amounts to 21 square feet. The bedroom door is 82 inches high and 36 inches wide, for another 24.6 square feet.

In all, the wall openings add up to 80 square feet. The total wall area, including openings, is 1760 square feet. You could order 1700 square feet and probably have enough lumber to complete the walls.

When you start on the walls, you can start at the top and work down or start at the bottom and work up. The advantage to starting at the top is that you will use partial boards, if necessary, at the bottom where they will be hidden by the baseboard molding. The disadvantage is that you must do the majority of the nailing upside down, which is difficult for most people.

You can measure the widths of boards and divide this number into the height of the wall to see if you will need a partial board. The height is 8 feet, or 96 inches. If you use 4-inch boards, you will need exactly 24 of them to reach the top. No partials will be needed. If you use 5-inch boards, you will need 19 boards with a fraction of an inch left over. The remainder, ⅙ inch, is negligible and will permit you to start at the bottom and work up, as will the 4-inch boards. Ceiling molding will cover such a small space if one is left.

Starting at the bottom is recommended because there will be no significant space to fill at the top because of the 8-foot walls. Before you nail anything, lay a level on the floor to be sure that the horizontal level is true. When the first board is positioned, check the level reading for the top of the board. Be sure to get started with a level reading. Do not begin with an off-level board because each board thereafter will be off level and you will have an awkward and unsightly space to fill at the top.

Begin with the two 12-foot boards. They will cover the 24-foot wall and fit perfectly. Alternate by using three 8-foot boards next. The three will cover the 24-foot space and leave no waste.

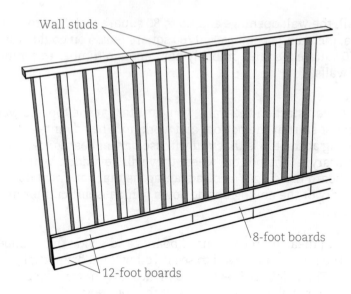

Wall studs

8-foot boards

12-foot boards

Fitting around windows

Continue this pattern until you reach the rough window opening. Here you may have to cut out the top part of the board to allow it to fit under the window comfortably.

Use 12-foot boards after the piece under the window has been fitted. You will have to waste about a foot at the end of each board when you start up the side of the rough window opening.

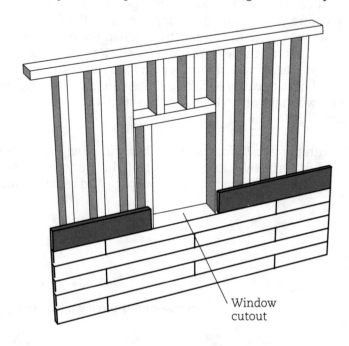

Window cutout

If you use a 10-foot board, you will not have enough to reach the window, and you will need to fit a 1-foot piece at the end of the 10-foot board. If you want to keep waste at a minimum, use a 12-foot board first, after cutting off 1 foot, then use a 10-foot board, add the foot-long piece you cut off to the 10-foot board, and you will have no waste.

On alternating courses add the 1-foot section at the window on one course; two courses later, under the 12-foot length, use the 1-foot length in the corner. The short length will not be noticeably odd if used this way.

Keep in mind that in 1992 tongue-and-groove boards were more than 50 cents per foot, and each time you waste 2 feet, you have wasted $1, which will add up to $15 per window, counting the waste on both sides. On a single floor, if the same pattern holds true, you would waste more than $150 in window fittings alone.

When the wall is complete, select the next worksite. If you choose to install boards on the bathroom-door wall, you will need to cover a 20-foot wall 8 feet high. The wall to the bathroom door is 17½ feet.

Installing around doorways

You can work out the least waste by using a 12-foot length board for the first part of a course. Then cut a second 12-foot board. Take 5½ feet from the board. Start the next course at the doorway and nail up a 12-foot board. Use the second half of the board you just cut, which is 6½ feet long.

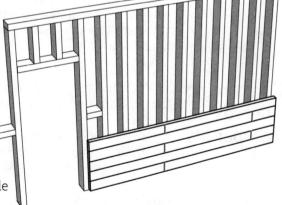

You will have 1 foot of waste. Stack the 1-foot sections in a convenient place and use them on the other side of the doorframe or behind the bedroom-hallway door where you will need 6-inch sections. Again, you will have no waste.

On the other exterior wall you will board around another window. Use two 10-foot boards for the first course. Do not repeat the same pattern for the next course. You must find a combination that equals 20 feet without having the boards end on the same stud. On the second course, start by using a 6-foot section, followed by a 10-foot section, followed by a 4-foot section. There will be no waste, and all cuts will end on a stud.

Then return to the two ten-foot boards for the next course and the three-board arrangement for the next course. Follow this pattern until you reach the bottom of the window.

If you used part of the wall space for a closet, your wall is only 18 feet long rather than 20. In this case you can use alternating courses of 12-foot and 6-foot boards. You will have no waste and will need to cut only one board for each course.

When you start up the sides of the window, you will have (if you use part of the wall space for a closet) 7 feet 3 inches on each side of the window. Waste at this point is unavoidable. Your best bet, perhaps, is to use 12-foot boards, which will leave 4 feet 9 inches. You can later use the 4-foot part of the board and still end on a stud for nailing.

Use the same type of figuring to arrive at lengths for all other walls in the house. Save all waste lengths, and you can find a way to use a remarkable number of them later, particularly if you install tongue-and-groove covering on the dormer windows upstairs.

Conforming boards for proper fits

When you are trying to fit a board and it refuses to conform to the previous one, use the leverage principle described in chapter 5. Use the C-clamp to hold a length of 2-x-4 solidly against a stud and above the board giving trouble. Use a short length of 2-x-4 or a pry bar as a lever and gently push the new board into position.

Hold the pressure on the board while you nail. You might have to start applying pressure at the end and work back toward the middle, or you might need to start in the middle and work toward the end. Several methods work equally well. If the first effort fails, try another. You can sometimes align boards well if you place a short board flat against the two boards being installed and strike the short board with a hammer.

At times you may need to use a pry bar to lever the top edge of the board installed so that it tilts out only slightly. At all times, use a scrap piece. Do not strike the board you are installing.

When you are butt-joining boards in the center of a wall, you should try to make the cuts (if cutting is required) as straight as you can. Even a tiny discrepancy will show up when the two boards are installed side by side.

Butt-joining methods for smoother installation

One trick when two boards must be butt-joined is to lay one faceup on the worktable and the other facedown on top of the first one. Align the end of one board with the saw mark or cut line of the other. When you saw, if you make a mistake, the same mistake will be made on the other board, but in reverse, and when the two boards are installed right-side up, the two ends will match perfectly.

Do not be overly concerned about cuts that do not match exactly in corners and at windows and doors. Door and window facing will cover most mistakes, and corner molding will cover the rest.

CHAPTER 8

More about ceilings

EARLIER YOU were given suggestions on how to install drywall ceilings and tongue-and-groove ceilings. This chapter is devoted to other ceiling coverings.

Ceiling tiles, among the most common coverings, are easy to install. The results are satisfying, and there is no complication involved. No expensive equipment is needed.

Selecting the tile

Return to the 20-x-24-room example discussed previously. If you decide to run tiles along the 24-foot length of the room, you will find that all basic dimensions of tiles will fit without cutting. You will need 24 12-x-12-inch tiles per course, 14 24-x-24-inch tiles, 18 16-x-32 inch tiles (installed with 16-inch running lengths across the room); 9 of the same tiles turn the other way.

On the 20-foot side, you will need 20 12-x-12-inch tiles per course, 10 24-x-24-inch tiles, and 15 16-x-32-inch tiles (turned the 16-inch direction). If you choose to run the tiles the 32-inch direction, you will need to cut the first and last tiles of each course.

Although the tiles are not hard to cut, it is obviously easier to install tiles that need not be cut. There is always the danger that you will mark wrong, cut wrong, or otherwise damage the tile. If you are equally happy with the other tiles, you might be wise to choose one that will fit without cutting.

Preparing the work surface

Before you start to work, you need to install *lath*, or strips, across the joists. If you prefer a thicker, more rigid ceiling, you can use plywood to cover the joist area completely and eliminate the lath.

If you wish to save the cost of plywood, you can buy 1-x-2, 1-x-3, or even 1-x-4 stock from which to cut the lath or strips. The tiles are very light, and you will not need very strong lumber for the lath.

Assuming that you choose 1-x-2-inch lath, you will install these against the first wall of the room. Then measure out the distance from the first strip to the point where the tiles will end and halve that distance.

You want to alternate or stagger lengths of the tiles for a better and more attractive appearance, so if you are using 12-x-12 tiles, start with half a tile at the beginning of every other course. Conclude the course with the other end of the cut tile.

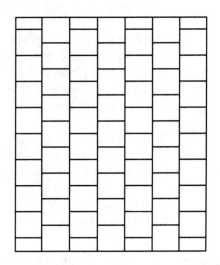

You can also install the tiles evenly—all four corners meeting without a staggered effect. By doing so, you will not need to use as many lath (the 1-x-2 or 1-x-3 strips recommended).

When the furring strips of lath are installed, double-check to see that the bottoms of all strips are exactly even. If the bottoms of the strips are not even, the tile cannot be even. Install the strips, then, 6 inches apart if you plan to use a staggered effect.

Start with a full tile and fasten it to the lath. Many builders prefer staples to nailing, and you will see a flangelike border on the tiles where you can staple.

Note that the tiles are manufactured with a tongue-and-groove edge so that when the first tile is installed, the second will fit into the first one neatly. The tongue-and-groove design will hold the end of the tile in place so that you do not need to staple on that edge. Staple only on flange edges.

Installing ceiling tiles

Your first tile will be started either in one corner or in the center of the room. If you plan to start in the corner, measure the room to be sure that you can use all tiles full-size. If you cannot, divide the first tile and use the other half at the end of the course.

If a room is 21 feet across, cut one tile down the center. Use the 6 inches at each end of the course.

The first tile is set in place so that the flange edges face the outer part of the room. Staple the flanges to the lath. Set the next tile into place and push the tongue-and-groove joints together. Staple the flanges to the strips. Continue in this manner across the room.

Hold the tile firmly against the strips while you staple. Use three or four staples on each flange. The longer the tile, the more staples you should use. Ask your dealer about the best staples.

If you choose to install plywood across the entire ceiling, you can cement the tiles in place rather than stapling.

Using adhesives for installation

To cement tiles, place a small amount of cement or adhesive (about a tablespoon) on each corner and 2 inches or so from each edge. When you set the tile in place, push it against the backing (plywood or strips) and move it to and fro slightly to spread the adhesive. Then push it into position and hold it for a few seconds until it adheres.

For larger tiles, install strips in accordance with the size of the tiles. Staple as before. Use more staples for the larger tiles, as many as eight per side on the larger sizes.

Before installing tiles in a room that is irregular (corners not square, for instance), use a chalk line to determine the correct square for the room and install along the chalk line to keep the pattern aligned and neat. Ceiling molding will cover a small discrepancy where the tiles fail to meet the walls.

Installing from ceiling midpoint

When you want to start in the middle of the room, measure carefully to see how many tiles will be needed. Find the midpoint of the wall and start with the edge of one tile on the chalk line that divides the room in half. Continue to install tiles the remainder of the way to the corner. Return to the midpoint and work to the opposite edge.

You can also find the exact center of the room and install in all four directions from the center point and centerline. To mark the centerline, measure the length of the room and mark the point on the ceiling joist on each side of the room. Measure the other direction and mark on both sides of the room again in the same way. You now have the midpoints clearly marked. Strike a chalk line in both directions. Where the two lines cross is the center of the room. Start installing tiles by setting one tile so that the tongue edges align with the two marks adjacent to it.

Suspending ceilings

Suspended ceilings can be installed in two ways. You can lower the ceiling by installing strips around the wall (a type of molding), then installing metal *runners* on which the tiles will lie, or you can do it the old-fashioned way.

Using runners

Measure down from the joist edge to the height you want the ceiling lowered. Do the same along all four walls, once at each corner. Strike a chalk line, then use your level to determine that the lines are correct. Nail up molding so that the bottom edge aligns with the chalk line.

Determine the position of the runners (in accordance with the size of the tiles) and install the screw-eyes in the joists. You will wire the runners to the screw-eyes. Keep a constant check on the level readings to be sure your ceiling remains constant horizontally.

The cross tees are then installed at right angles to the runners. Follow manufacturers' instructions.

When all the framework is installed, you are ready to lay the tiles between the runners so that the tiles rest on the ledges on the inside edges of the runners.

Lowering ceilings

To lower a ceiling or suspend it the old-fashioned way, determine the amount the ceiling is to be lowered and measure down from joists, as you did when preparing to install suspended ceiling, and mark the level on the wall. Strike chalk lines as before.

Fasten 2-x-4s with the bottom edge aligned with the chalk marks. The 2-x-4s should extend completely around the wall. Measure and mark the points where you want the tiles to be installed. If you are using 24-inch tiles, lay off the wall in 24-inch gradations on both opposite walls. Chalk lines across the bottom edges of joists over the entire room.

If you are lowering the ceiling 1 foot, cut some lengths of 2-x-4 16 inches long. Lap 4 inches on the joist edge and nail the 2-x-4s to the joists where the chalk marks are located.

When all the down-lengths are installed, you are ready to install the *nailers*, or runners. These can be 2-x-4s or 2-x-2s. If you need to save money, rip a 2-x-4 into 2-x-2 stock and nail these to the down-lengths.

Always nail the strips to the sides of the down-lengths. If you nail up into the bottoms, you always run the risk that the nails will pull loose and the ceiling will sag. Never let the end of one of the down-lengths hang even a fraction of an inch lower than the bottom edge of the runners you are installing. When you are cutting the 2-x-4 lengths, use a square to get a straight line and cut very carefully along the line. A slanted cut can produce serious problems later.

When the runners are in place, you are ready to install the tiles as you normally would. By lowering the ceiling or suspending it, you can cover unsightly pipes and ductwork, and you can save on heat bills because you have eliminated several square feet of space from the room. If you lower the ceiling of a 20-x-24 room, you have eliminated 480 cubic feet from your heating space and cost. This is equal to one-eighth of the cost of heating that room.

Other ceiling coverings

Ask your dealer about other types of ceiling covering. There are several other materials that are installed in essentially the same way as ceiling tiles.

You can cover drywall with compound used in taping drywall. This compound goes on over almost any type of solid surface such as plywood and similar materials.

You can buy a roller that is patterned or molded so that the compound is applied in designs rather than a smooth layer. This compound dries so that it is hard to the touch and will resist changes in heat and light without problems. It will not resist moisture or rough treatment. You need not paint the compound. It is a glistening white when it comes from the 5-gallon can, and it does not fade or lose its brilliance after application.

This is called a *textured ceiling*. You can also buy textured paint (some people mix their own sand into the paint to produce their own textured paint at a much lower cost than you find at supply houses). These textured surfaces will dress up the ceiling of nearly any room.

CHAPTER 9

Paneling & other wall covering

Y OU ARE NOW ready to look at methods of installing paneling and other forms of wall covering.

No book and certainly no chapter can include all the ways to cover walls. This chapter is intended to equip you to install one of the most popular forms of wall covering. You can translate these skills into the ability to use them for installing other materials.

When you buy, be sure that more of the same style of wall covering will be available if you need more. You can overbuy with the understanding that you can return for credit or refund any panels that are left over and undamaged.

Installing paneling

When you start to install paneling, lay a level on the floor to see that it has remained level. Place the level on the bottom edge of the joists of the ceiling to see that they too are level and the bottom edges are exactly the same height. Use a square in the corners to see whether the corners remained stable as you framed the house.

Begin installation in a corner, as you did with drywall. Most of the principles you learned in chapters 4 and 6 will apply to installing paneling. A few major differences are discussed in this chapter.

When you stand your first panel in the corner, note whether it fits well. If it does not, you must modify it. You must always start with a square corner fit, and if the corner is wrong, you must modify the other paneling because it will be too late to alter the corner framing.

With the paneling in position, drive a small nail into the last stud covered partially by the panel and against the outside edge of the paneling. The nail should hold the panel in place while you step back and examine the fit. Do not drive the nail through the paneling itself. If the fit is good, you can install the panel. You can buy colored nails that match the paneling and are not noticeable unless you are looking for them.

If the panel does not fit, measure the discrepancy and mark and cut accordingly. Remember that if the corner is not square—when the panel is held at a perfect horizontal position, there is a space at the top or bottom of the panel—you must cut from the opposite end to get the panel into a square fit.

Measuring and cutting for proper fit

If there is a space of 2 inches between the top of the paneling and the corner post, you must take down the panel, lay it on a work surface, measure 2 inches from the bottom, and mark the point. Then strike a chalk line from the bottom mark to the top corner. When you cut off this slender triangle, you will have the panel cut to fit the corner.

Chalk mark

When you cut paneling, always lay it facedown and cut with the back side up. Circular saws and jigsaws will splinter on the top side of the material being cut, so if you cut the panel upside down, the splintered area will not show.

Stand the tailored panel into place and hold it firmly against the corner post while you drive at least two or three nails partly in. Put a level on the outside edge of the panel once it is in place. You should have a true vertical reading. If you do not, further adjustments must be made.

Continue to install other panels as you would install drywall. If you must cut a panel, cut from the side that will be butted into the corner. Leave the factory edge to butt into the panel already installed.

Installing around receptacles and switch boxes

Cut out for receptacle and switch boxes as you move around the room. Measure from the floor to the bottom of the receptacle, then to the top. Write down both measurements. Then measure from the edge of the paneling just installed to the near side of the box, and then to the far side.

You now have your parameters established. Lay the panel on its face and measure up from the bottom to the bottom line of the receptacle box, then from the top of the box. Use a square to help you mark straight lines across the back of the panel. Now measure over from the edge of the paneling the distance to the near side of the box, then to the far side. Use the square to draw vertical marks showing both sides of the receptacle or switch box.

You now have a rectangle formed by the marks—the outline of the box. Cut out the rectangle, and the panel should fit well.

To cut out the square, you can use a drill bit positioned so that the outside of the cutting edge will barely touch the two lines in one corner of the rectangle. Drill until only the point of the bit is through the wood. Do the same at the other corners. When four tiny holes are made, turn the panel over and drill from the face side. Place the point of the drill into the holes already made and drill the rest of the way through the panel.

Remember that drill bits splinter the wood when they emerge, so if you turn the panel faceup, the splintering, occurs on the back side of the panel.

When all four holes are bored, used a jigsaw or keyhole saw to cut out the rest of the rectangle. When the panel is placed in position, the fit should be good. If it is off slightly, you can trim the edge of the cutout to secure a better fit. It does not matter if the fit is slightly loose. The switch plate or receptacle covers will hide small problems.

Paneling around windows and doors

When you must cut out for windows and doors, measure and mark the cutline and use either a circular saw or a jigsaw to make the cuts. To measure, start at the edge of the last panel installed and measure 4 feet away, so that the mark is on the studding over and below the window. Then measure the distance from the mark back to the edge of the window framing.

You will need to cut out that amount of paneling. Measure the length or height of the window as well before you cut.

When you must cut paneling, always leave the factory edge to butt into the last panel. If you must cut a small piece to fit between installed panels and you cannot use both factory edges, cut along the grooves in the panel. Cut all the way to one side or the other. Do not try to cut down the center of the groove.

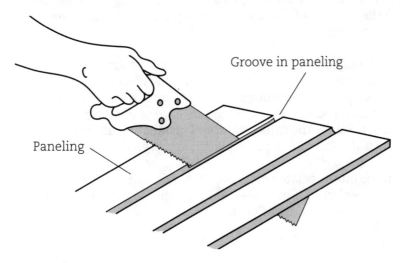

Groove in paneling

Paneling

Paneling & other wall covering   79

Installing solid-lumber paneling

If you do not want to use panels, you may wish to use solid-lumber paneling. You can buy a wide variety of widths and types of wood for this purpose. For all walls, the V-joint tongue-and-groove boards in alternating widths work well for a highly attractive wall.

You can install the boards in three ways: horizontally, vertically, and diagonally. In another way, you can mix the three methods and have part of the wall vertical with the rest of it diagonal, or you can come up with several combinations to your own preference.

The easiest type of installation is horizontal. You can nail to the studding. Use cut nails and install as you were directed in chapter 7.

If you install vertically, there is nothing between the studs for you to nail to, so you must install nailers, sometimes called furring strips. You must have at least three nailers: one at the very top of the wall, one at the very bottom, and one at the center. For a much stronger wall, use furring strips every 1 or 1½ feet. You can use 1-x-3 nailers for lighter boards. The only minor difficulty is that in the center of the space between studding the 1-x-3 board tends to give slightly as you nail.

For heavier boards (and to keep from losing any room space) you can install 2-x-4 lengths between studding. Arrange them in the manner of the old bridging used before builders began to cut too many corners.

At the top, nail 2-x-4 *bridging* between the studs and to the top plate. You can also nail through the studs (except for the corner posts and partition posts) and into the ends of the 2-x-4 sections. You can also use the side edge of the top plate as a nailer. Follow the same pattern of installation at the bottom of the wall. Nail into the sole plate if you prefer.

In the intermediate portions of the wall frame, cut the lengths for a perfect fit if possible. Nail through the side of the stud and into the ends of the bridgers or nailers.

For easier nailing, stagger the nailers. Measure up from the floor 18 inches (or whatever spacing you prefer) on each end of the wall. Mark the points, then strike a chalk line between the two marks so that you can read the desired height on the outside edges of the studding.

Install a nailer between alternate pairs of studs. You can nail easily through the studding. For the remainder of the studding, install the nailer either above or below the nailer you just installed. Let the bottom edge of the new nailers align with the top edge of the ones already installed. If you want to nail them in below, let the top edge of the new nailer align with the bottom edge of those already installed.

When nailers are in place, nail up the boards. You can angle-nail all boards (except those that are extremely wide) through the tongue and into the studding, and the nail heads will not show.

If you buy end-matched boards, you can use odd lengths at various places across the wall surface. The use of short pieces for a varied effect is attractive, more so in many respects than using all 8-foot boards.

If you install diagonally, you will need to strike a chalk mark from one top corner to the opposite bottom corner. This mark divides the room equally. Measure from the mark at two or more points to get the right angle for the corner lengths.

Diagonal installation

If you are using 5-inch boards (or you plan to use one in the corner) measure to the midpoint of the diagonal line from corner to corner. Mark the point. Then strike a chalk line from that point to the corner of the wall.

Use your square (a small instrument like the quick square is more useful here because it requires less room), hold one side of the square on the line to the corner, and mark along the other side of the square. Do this on both sides. You now have the proper line to use for the starting point.

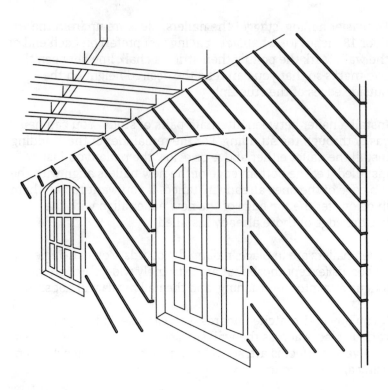

If you do not mark the corner properly, your angle may be off so badly that when you reach the opposite corner, the entire wall will have an unbalanced look.

If you become confused by the marking, there is another way that is very simple and impossible to confuse. Lay a short length of board so that one edge parallels and aligns with the chalk line. Mark along the outer edge of the board. Then move the board so that one edge parallels and aligns with the latest line made and mark across the other edge of the board.

Continue this pattern for the remainder of the space. It will take only a few minutes, and the results are never in doubt. You can nail to studs and eliminate the need for nailers. You will have a slight difficulty in installing the final short lengths in corners. If you cannot push the final board into place, you can trim the tongue slightly (or completely off) to get the fit you need.

You can also buy plywood panels manufactured with grooves or simulated wood grains. These are installed exactly like paneling. *Hardboard*, made from wood fibers, comes in a variety of patterns and can make an attractive wall.

When you are using any wall covering that contracts and expands (as wood does) when temperatures and humidity readings fluctuate considerably, unwrap the bundles and lay them out so they can acclimate to room temperature.

There are other wall coverings available. When you are ready to cover walls, consult with your dealer. He or she can tell you what the latest items on the market are.

You can also return to old-fashioned plaster (updated and far simpler to install than it was decades ago). Plaster has many advantages, among them beauty, variety of colors through painting, and sound insulation. Fire retardation is one of the finest qualities of any material made from gypsum, as plaster is.

Most do-it-yourselfers are not sufficiently experienced to apply plaster, so this material has not been treated extensively in this book. If you want plaster wall covering, your best bet is to install the plaster base yourself and call in a skilled workman to finish the plaster. You can buy a perforated gypsum base to nail over studding. You then mix the plaster according to manufacturer's instructions, spread it with a trowel or smoothing tool, and let it dry. This first coat is called the *scratch coat*.

You can buy or rent a plaster machine to apply a base coat. When the scratch coat is applied and you are ready for the final coat (the combined thickness of the three coats should be at least ½ inch), use a trowel and float to smooth the gypsum-lime mixture.

Smoothing the plaster requires skill, patience, and a thorough knowledge of plastering. Many old-time plasterers use a flashlight held flat against the wall to detect uneven places. The shadows cast by the light reveal trouble spots.

Other wall coverings

CHAPTER 10

Completing stairways

ONLY AFTER the wall coverings are in place should you begin work on completing stairways. When the walls are complete, you can begin the work of installing finish treads, balusters, or rails.

You may have read that stairways are among the most difficult and intricate elements of house construction. This is true if you are concerned with elaborate and highly expensive stairways. If you are interested only in an attractive and useful set of steps, you can complete the work yourself.

You can order predesigned and premanufactured stairway parts, or you can have the entire unit delivered and installed, depending upon the complexity of the work. In 1990 we asked about a stairway we had seen in a showroom and were told that it would cost about $25,000 for the stairs. At that point we decided to build our own. We built an entire 4300 square-foot house for less money than the stairway would have cost, and it seemed pointless to demolish the budget at this point.

Installing treads

You have already cut stringers, or the carriage, and installed these, and you have already nailed temporary treads for use while the heavy building work was under way. You can now install the final treads and risers and build your railing and baluster units.

You can order factory-made treads that will work well on your stairs. These treads are available in a variety of widths and lengths and are not extremely expensive. If you prefer, you can buy wide boards and cut your own treads. You can buy the ready-to-use treads for about the same price the boards will cost you.

Treads are often 12 inches wide (deep) and 3 feet long. If you need less width, you can use the circular saw and rip off part of the back of the unit. Do not rip or cut off the front because this is the rounded part that you want for a slight overhang.

Do-it-yourselfers are advised to install only straight-run stairs that lead directly from one floor to another. Winding stairways are far more difficult to build and cost a great deal more for materials. They also require much more room than the typical straight-run stairs.

You can install *treads* and *risers* in two slightly different ways. The first is to nail in the treads and later, when time and economy permit, install the risers so that the bottom edge sits on top of the tread and the top edge fits under the nose of the tread.

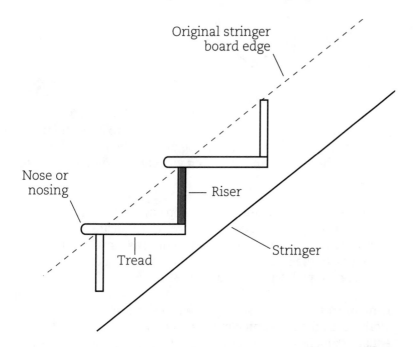

The second and preferred way is to install the risers first, then let the bottom tread butt into the face of the riser; the top tread can rest on the top edge of the riser. One advantage of the latter

Installing risers

method is that you get greater width from the treads if they are not backed all the way into the cutout for the stringers.

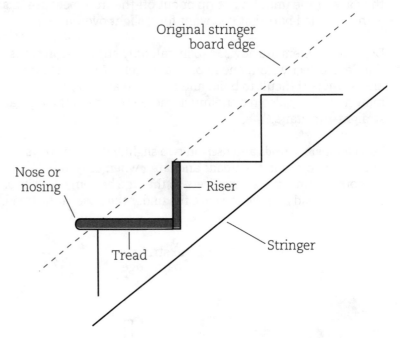

Original stringer board edge

Nose or nosing

— Riser

Tread

Stringer

Another advantage is that the riser will help to hold the treads securely in place. Stairways can be among the most hazardous places in the house, and anything you can do to increase safety is money advisedly spent.

To install risers, measure from the top edge of the stringer to the corner forming the nest tread position. If this distance is 7½ inches, buy an 8-inch board. You will find, as usual, that the finish size of the lumber is less than the nominal dimensions. An 8-inch board will usually measure 7½ inches.

Remember that the treads and risers of stairs should work out so that the unit run plus unit rise should equal 17½ inches. The *unit run* is the length of the cutouts for the treads, and the *unit rise* is the distance from the top of one tread cutout to the corner of the one above it.

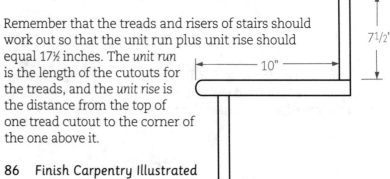

7 1/2"

10"

If your treads are 10 inches, your risers should be 7½ inches. You can install risers behind the treads if you need to, to work out the basics of the stairway construction.

Another way to examine the problem is that the height of two risers and the width of one tread should equal 25 inches. If your risers are 7½ inches, two of them will equal 15 inches. Your tread, then, should be 10 inches wide. Remember that if you install your tread first and let the riser sit on the top edge of the tread, your 10-inch tread becomes a 9-inch tread. The two risers then should equal 16 (25–9) inches. Each riser should then be 8 inches.

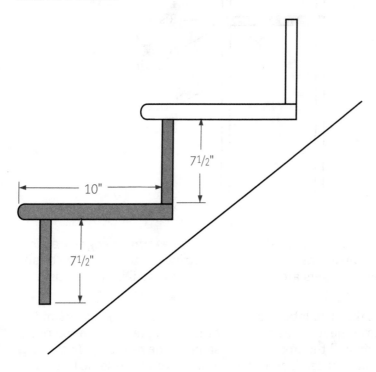

The most important thing to remember is that you should make all treads the same width and all risers the same height. If you and the building inspector can agree on riser height and tread width, dimensions that suit your family and stairway users, you can manipulate the figures slightly.

Cut the risers so that they will extend an inch past the outside edge of the stringers, or an inch past the wall covering if you plan to install molding around the treads and risers. Use 3-inch finishing nails to fasten the risers to the stringers. Use at least three nails in each stringer.

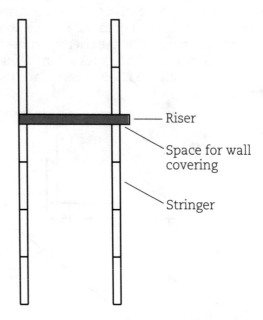

Double-check to see that no part of the stringer edge extends upward past the level line of the stringer cutouts for the treads. When the risers are all in place, you can nail in the treads.

One-step installation

To make installation easier, you can start at the bottom and remove the temporary tread for the first step. Nail in the riser, then install the tread. Do the same at the next step. In this fashion you can install risers and treads at the same time, and you will have a place to sit or kneel while you work.

Fit the treads into their positions and nail them to the stringers with three finishing nails for each stringer member. The treads should extend an inch past the outside edge of the riser. This slight protrusion is sometimes called *nosing* or a *noser*, as you will remember.

Your next step is to install the balusters and railing. If economy and simplicity of design appeal to you, you can make a very attractive stairway rail and baluster combination by using 2-x-4s angle-cut and doweled.

Decide first how high you want the railing to be. A good workable height is 36 inches from the top of the tread to the top of the railing. You will find that the point where the baluster is to be installed will cause the length to vary slightly because of the slant of the railing.

A simple plan is to cut a length of 2-x-4 that is 36 inches long. Determine the best-looking side and turn it so that it faces upward. Measure down three inches on one side and mark the point. Use a straightedge to mark the line from the point just made to the point created by the end cut of the 2-x-4.

Cut the 2-x-4 at this line. You now have a length that is 36 inches long on the high side and 33 inches long on the low side. Cut another length exactly like it and nail the two lengths together so that the cut lines match and the edges are aligned perfectly all around.

This doubled 2-x-4 section will now become the final rail post at the bottom of the stairs. Install the rail post by using dowels and wood glue for maximum strength and resistance.

To install dowels, lay the combined unit on a work surface and clamp it in place. Mark the exact center of each 2-x-4 length. Measure from side to side and end to end to get the exact center. Mark end points. Decide at this time how large the dowels will be. You can buy 3-foot lengths of dowel at most supply houses, and these can be from ¼ inch to 2 or 3 inches in diameter.

Half-inch dowels work well. The 2-x-4s are 1½ inches thick and 3½ inches wide. Use a ½-inch drill bit and drill a hole starting at the midpoint mark you made and 3 inches into the 2-x-4. The hole will leave ½ inch on each side of the hole on the narrow side and 1½ inches on the wider side. Be sure to hold the

Planning baluster and railing installation

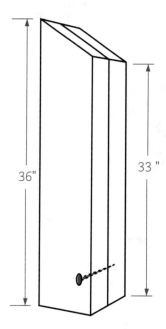

36"   33 "

Using dowels and wood glue

drill level and straight, so that the hole will be drilled straight into the ends of the 2-x-4s.

Cut lengths of dowels to reach all the way into the holes and leave at least 2 to 3 inches sticking out. Remove the dowels and coat the upper 3 inches (the depth of the hole) with wood glue. Reinsert the dowels.

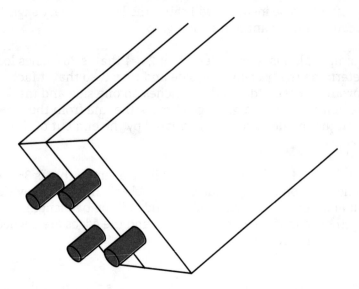

Now stand the 2-x-4 assembly on the desired position on the bottom tread. Use a level to be sure that you are holding the assembly perfectly vertical.

Using a pencil, reach under the baluster unit and mark around the ends of the dowels. Remove the baluster assembly for the moment and drill a 1-inch hole down through the tread and into the riser top. This hole should also be 2 to 3 inches deep.

Clear away all sawdust or shavings from the holes and use a brush to coat the sides of the hole with wood glue. You can use the squirt tip of the glue bottle to direct the glue properly.

Set the dowels into the holes and push them down firmly until they are seated at the bottom of the holes. Remember that the upper or longer side of the assembly should be toward the upper part of the stairway.

Use the level again to position the assembly so that it is exactly vertical. Check both sides to be sure that it is not leaning in either direction. If you wish, you can toenail finishing nails through the sides of the assembly and into the treads.

Lay a long 2-x-4, wide side down, across the top ends of the installed assembly, holding the 2-x-4 so that the bottom side of it fits the slant of the installed assembly perfectly. Have a helper hold the 2-x-4 in position while you position a single 2-x-4, narrow edge against the 2-x-4 held in place, and get a level reading true vertical. Mark along the underside of the railing 2-x-4 and across the face of the 2-x-4 being held vertically.

Measuring and installing hand rails

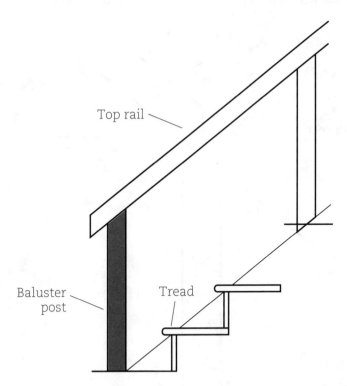

Top rail

Baluster post

Tread

Cut the held 2-x-4 along the marked line. Cut another one exactly like it and assemble the two pieces as you did before.

One 2-x-4 is not likely to be long enough to reach more than partway up the stairway. Plan to use two lengths joined at the midpoint of the stairs.

Install the doubled baluster exactly halfway up the stairs. Use dowels and glue as before.

Now position the railing 2-x-4 flat on the two installed assemblies. Have a helper hold it so that the top ends stop halfway across the unit installed there. Use a level to mark the vertical cut line of the railing.

Hold the railing in position and go to the bottom assembly. Measure out 2 inches from the far edge of the installed baluster assembly. Mark on the bottom of the 2-x-4 railing, then use the level to mark a vertical cut line straight up from the bottom mark.

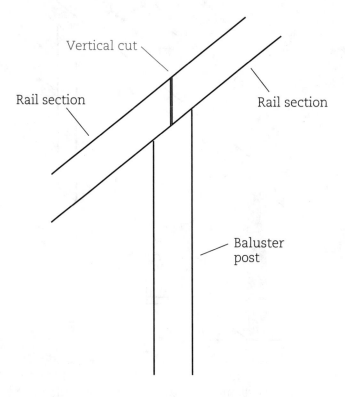

Vertical cut

Rail section

Rail section

Baluster post

Take the railing down and saw along the two cut lines. You can now install the railing by using finishing nails or dowels. If you use dowels, drill only an inch into the bottom side of the railing. Mark the drill points by holding the railing in position, then making a light mark on each side of the railing at midpoint where the railing crosses the doubled balusters. Drill the 1-inch holes and install the dowels.

Hold the railing in place and mark for drill holes in the top ends of the balusters. Drill, install glued dowels, and place the railing in position. If you use dowels, no nails will show. If you choose to use nails, you can use a finishing nail and use a punch to countersink it so that the head will not be visible.

Now install single balusters, two per tread, between the two double balusters. Stand 2-x-4s on end and use a level to get a true vertical reading. Mark under the railing and across the edge of the 2-x-4.

Completing balusters

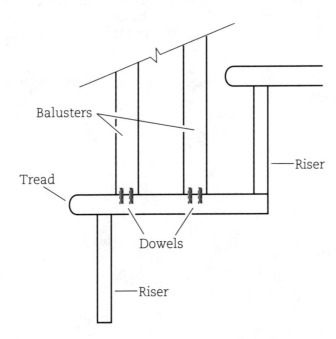

Hold the 2-x-4 so that the narrow edge is against the railing. When the baluster is installed, it will be as wide as the railing, and you will have double support because you installed two balusters per tread.

To determine the positions of balusters, measure the top of the treads. Your balusters should be located so that the first one is 1 inch back from the corner of the riser and tread on the stringer and the other is two-thirds of the way to the riser at the back of the step.

When the section is complete, go to the top of the stairs and hold a length of 2-x-4 with one end cut to match the end of the first railing piece. Hold the free end so that it aligns with the slant of the other section of railing. Measure to the top of the treads and mark as you did before.

Install the double baluster at the top of the stairway, then cut the railing section so that one end fits neatly into the space left at the midpoint double baluster. The other end should be flush with the outside edge of the double baluster at the top of the stairs. If you prefer, you can allow the railing to project 3 or 4 inches.

Install all balusters in the second section. The stairway is now complete.

Installing
deck railing

DURING THE rough-carpentry stages of deck building, you left the deck area when it had been floored. You are now ready to install the railing and, if you choose, steps.

Your decision should take into consideration how much your deck lends itself to privacy. If the deck is elevated, you may want to omit steps to assure yourself of a relaxation area that is protected from stray animals, trespassers, and other inconveniences.

To install a full railing (with no steps), you should first check with the building code. Probably you will need to use treated lumber because the wood will be exposed to the weather.

You have two logical points at which to start: at either corner of the deck adjacent to the house. The deck designed in *Rough Carpentry, Second Edition* (TAB #4395) was 10 feet wide and 52 feet long. At this length, the deck will cover the entire front of the house. You can make it as small as you wish, but the full-length deck is not a great deal more expensive than a partial deck. In 1990 when we built our deck, the total cost of the entire deck, which includes more than 500 square feet, was less than $900, and that includes all of the concrete work, nails, and all other materials.

By comparison, at that time, local deck builders were asking more than $7500 to build a similar deck. Keep in mind that you will need relatively little more materials for the full deck than you will need for a half or three-quarter deck.

Start at the house side of the deck (either side of the deck will work equally well) and install a deck-railing post so that it is bolted to the headers of the deck and nailed to the house. If

Installing deck-railing posts

your flooring extended past the outside edge of the header slightly, use a handsaw to cut back from the outside edge to the header on two sides, then use a chisel to cut away the third side of the cutout.

When you are ready, choose a double-2-x-4 assembly for the first post. Let the first 2-x-4 of the post reach from a point flush with the bottom edge of the header to a point 34 or 36 inches above the floor level.

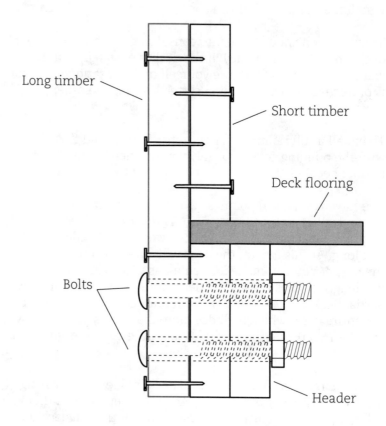

Cut another 2-x-4 and let it reach from the floor to a point flush with the top edge of the first 2-x-4. Nail these two units together or bolt them securely.

Stand the finished assembly so that the short 2-x-4 sits on top of the flooring and the longer length reaches down the side of the header and is flush with the bottom edge. Drive two 16-penny nails through the side of the 2-x-4 and into the header to hold it securely in place while you finish installation.

From the patio below, stand on a short ladder and drill two ½-inch holes through the 2-x-4 and the header. You will need ½-inch bolts at least 5 inches long to reach through the header and post. Let the round head of each bolt remain visible on the outside of the post. On the inside, attach a washer and bolt and tighten securely.

Use a wrench to tighten each bolt until the post is held firmly in place. When you go back to the deck, you can toenail through the end of the shorter post and into the deck flooring on at least two sides.

Building corner posts

At the corner away from the house, install a corner-post assembly. Make this by nailing or bolting two 2-x-4s together as before. One of the 2-x-4s should reach from the bottom of the header to a point 34 or 36 inches (the choice is up to you) above the floor of the deck. The other should reach from the top of the flooring to a point flush with the top of the first 2-x-4.

The assembly post works best if you install the short timber against the long timber that will be installed on the front or long part of the deck. Such a method will permit you to install the top railing better when you are ready.

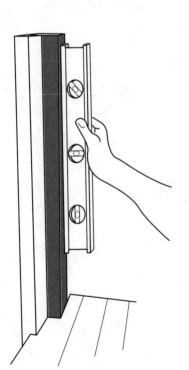

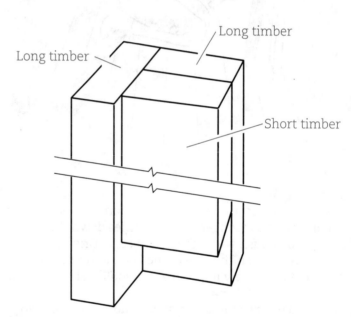

Long timber

Long timber

Short timber

Install the assembly as you did the first one. This assembly should be located so that one edge of the longer 2-x-4 is perfectly flush with the corner of the deck headers. Nail, then bolt.

Cut another 2-x-4 long enough to match the longer of the two. Install this second 2-x-4 so that it laps or covers the narrow edge of the 2-x-4 already installed.

You now have a corner-post assembly that consists of two long 2-x-4s securely attached to each side of the header corner and one shorter 2-x-4 firmly nailed or bolted to one of the long timbers. The end of the short 2-x-4 rests on top of the deck flooring.

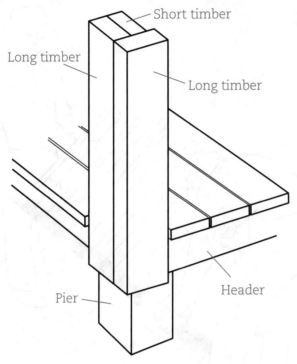

Short timber

Long timber

Long timber

Pier

Header

Nail the second long timber to the other long timber by using 16-penny nails spaced a foot apart. Then toenail the short timber to the deck flooring.

At this point you have the corner-post assembly attached at every necessary point. Drill bolt holes and tighten washers and nuts over the bolts.

It is now time to construct the baluster-and-rail assembly. You can do this on the floor, then lift it into place. You can also install the two rails and install the balusters between the rails after the rails are in place.

Constructing baluster-and-rail assemblies

For ease of nailing and accuracy of installation, you can nail the assembly together on the floor, then set it in place. An even better way is to nail only the bottoms of the balusters, then nail the tops after the rails are attached.

Bottom

Measure from the back side (the one adjacent to the house wall) of one post to the outside edge of the corner post installed. This distance represents the length of the top rail. Then measure from the inside edges of the two posts and cut a rail that length. This is your bottom rail.

Your balusters should be short enough that when the section is nailed in place, the bottom rail is three inches or so off the floor. Do not nail the rails to the flooring because moisture will accumulate under them and hasten deterioration.

If your top rail is 1½ inches thick (the usual thickness of a 2-x-4), the lower rail is the same, and you have 3½ inches of space left under the bottom rail, your balusters will not be 36 inches long. They will be 36 inches less the thickness of each rail and the height of the space under the bottom rail (1.5 plus 1.5 plus 3.5 inches equals 6.5 inches; this length subtracted from 36 equals 29.5 inches, the length of the balusters).

Your next problem is deciding how far apart to set the balusters to have a neat and symmetrical appearance in the railing. If you deduct the thickness of the corner post (3½ inches) and the thickness of the first post (3 inches), you have 6½ inches to take from the total length of the division for balusters. Your bottom rail is 113½ inches long.

The easiest way to determine spacing is to locate the exact midpoint of the bottom rail, which will be at 56¾ inches. Measure 56¾ inches from the inside edge of either post and mark the point. Your first baluster will be installed here. Now measure the distance between the inside edge of the baluster and the inside edge of the corner post. Divide this distance by 2 and mark the point. The second baluster will be located here.

Continue dividing the distances until you have the balusters spaced as you want them. Spacing is a matter of preference, but you do not want spaces that would allow a child's head to be caught between balusters or items used on the deck to bounce or blow between them and into the yard. A good spacing to try is 5 to 7 inches on center. Work out your best solution by dividing the distance until you have what appeals to you.

Turn the bottom rail upside down and mark the baluster placement on the bottom. Cut several balusters, all the same length, and install these by holding them upright under the bottom rail and nailing through the rail into the ends of the balusters.

Spacing balusters

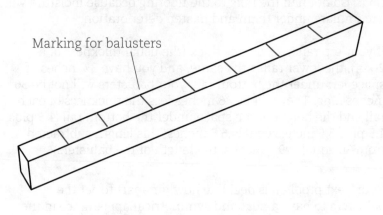

Marking for balusters

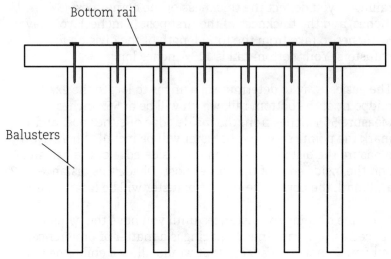

Bottom rail

Balusters

When all the balusters have been installed, turn the assembly over to its proper position and lay the top rail over the balusters. Be sure that you do not align the ends of the rails at one end of the assembly. Allow the top rail to extend 3½ inches past the ends of the bottom rail.

If you wish, you can install the top rail in its correct position on top of the corner posts. The ends of the rail should extend to the outside edges of the corner posts. At the outside corner, align the outside edge of the rail with the outside edges of the

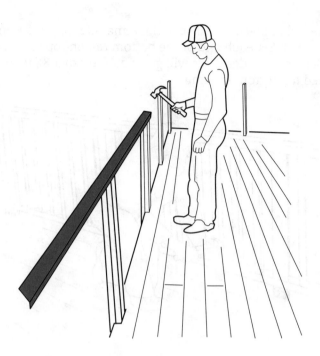

corner-post assembly. By doing so you will leave space for the rail along the front of the deck to seat with a good nailing and support space.

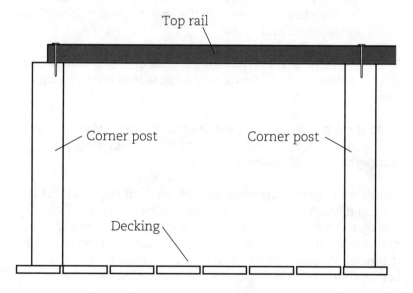

Top rail

Corner post   Corner post

Decking

Nail the rail in place. Then slide the remainder of the assembly into position. Set each end of the bottom rail on top of a block of 2-x-4 stood on edge. This will give you the exact 3½ inches you need for bottom spacing.

Hold a level to the side of the first baluster. When it is perfectly vertical, nail it in place by driving 16-penny nails through the top of the top rail. Continue across the entire length of the section in the same manner. Use the level to get a true vertical reading before you nail.

When you have finished, toenail the ends of the bottom rail to the corner posts. You now have one full side of the rail built and ready for use.

Go to the other end and repeat the process exactly as you did on the first end. You now need to do only the long expanse across the front of the deck.

Measure the remaining distance and mark off the locations for the rail posts. You have 52 feet less 6 inches, and you divide this by the number of posts you intend to install. Posts should not be more than 6 or 8 feet apart for maximum holding power. If you set the posts 6½ feet apart, you will need seven posts between the two corner posts already installed. Because you

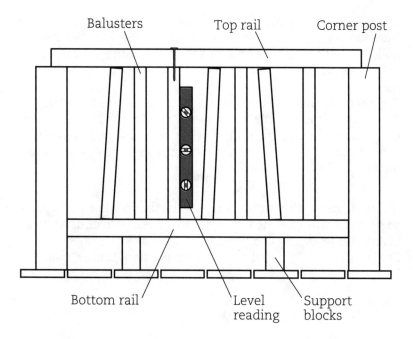

Balusters   Top rail   Corner post

Bottom rail   Level reading   Support blocks

used 3 inches total on the short 2-x-4s installed at the corner posts, you will need to make a minor adjustment if you want exact spacing. Make the spacing 6 feet 5½ inches if you need to be precise. The tiny discrepancy of the first method will not be noticeable, and you can keep the 6½-foot spacing if you wish.

Lay off the spacing for the posts. Construct posts as you did the first ones, with one long and one short timber. The two are nailed together, and the long one is bolted to the header, the short one toenailed to the decking.

For top rails, once the posts are in place, measure from the corner post (you will have 2 inches or so of nailing space on each corner post) to the midpoint of the second post. Always leave nailing space for the rail that butts into the one you are installing.

Using long timbers

If you are using 12-foot timbers, set the first posts 6 feet apart, so that you will have no waste. Do this from both ends and install full-length 12-foot timbers as long as you can. Set the final posts slightly closer together for the final timber span.

Install balusters as you did before. Nail them to the bottom rail, then set the unit in place and use a level to get a true vertical position before nailing them to the top rail.

You have not left a gate for a stairway if you plan to install one. If you need stairs or steps, you can frame a stairway opening into the deck flooring at either end, as you framed the stairway opening in the house.

You can also build a small landing outside the deck floor and build piers or use treated timbers (no smaller than 6-x-63) to hold the landing. Brace the timbers well and be sure they are plumb before beginning work.

Building a landing

The landing should be at least 3 feet by 3 feet, and the steps should be constructed in accordance with the methods suggested in chapter 10. Use treated lumber at all points where wood is needed. At the bottom of the steps, pour a concrete pad or footing for the steps to rest upon.

If you choose to frame the stairway opening in the deck area, take out the final joist before you begin to floor the deck. Leave the headers in place but remove the joist. You will have less than the recommended 36 inches for a stairway but an adequate width, slightly less than 32 inches.

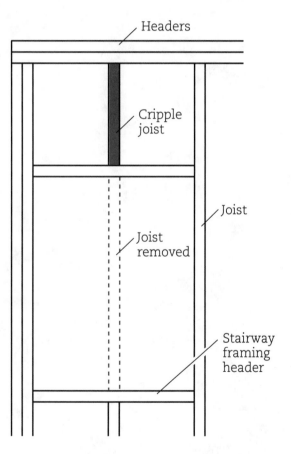

Headers

Cripple joist

Joist

Joist removed

Stairway framing header

Lay out and cut stringers as you did before. Install the stairway header as instructed earlier and install the stringers. When the steps are complete, you will need to install railing so that the steps are safe. Install this railing on the floor level as well as on the stairs themselves.

Installing
kitchen cabinets

YOU CAN MAKE your own cabinets if you wish to spend the time, money, and energy required to do so. It is false economy to attempt cabinetmaking unless you are well equipped to do the work and you have expertise in the area.

You can buy your cabinets ready-made for a reasonably low price. By the time you figure your materials, your time, and the travel to buy the materials, you have a considerable sum of money tied up in the work. Cabinet materials are very expensive unless you want to use one of the basic plywood exteriors, which do not yield the satisfactory results you might expect.

You can try your hand at making your own cabinets, and if your results are pleasing, you have saved at least some money, and your personal satisfaction is perhaps worth as much as the money involved. A few basic guidelines should be observed in virtually all types of cabinet work.

Units that sit on the floor, sometimes called *base units*, are typically 3 feet high. You can modify this height for your own needs if you are extremely short or tall, but keep in mind that extreme heights may interfere with selling the house if you relocate later.

Before starting to work, check with local building-supply houses to get prices on materials. Each panel is expensive, and you need to plan your work carefully before you begin cutting. The old adage about "measuring twice and cutting once" is especially relevant here.

Of the two ways to construct cabinets, one is preferable to many carpenters. Although some like to construct the units by building them into the wall, it is sometimes easier to build the

Basic methods of constructing cabinets

units in the middle of the floor, then hang them. Any modifications can be done easier with the unit on the floor than with it hanging from the wall.

Building base units

Make your base first for all units. Use sound, new, straight 2-x-4s to frame the base, which should be 2 feet wide; that is, the units should extend from the wall into the room a distance of 2 feet.

Assuming the unit will be 8 feet long, lay out two straight 2-x-4s and place them so that it is 2 feet from the outside edge of one 2-x-4 to the outside edge of the other. Measure and cut lengths of 2-x-4 to fit exactly between the two lengths already placed.

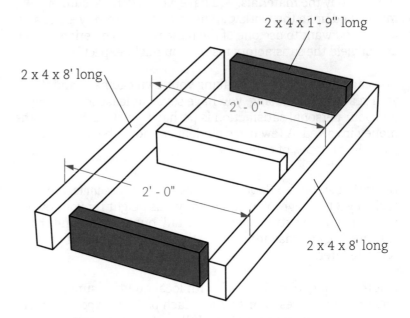

2 x 4 x 1'- 9" long

2 x 4 x 8' long

2' - 0"

2' - 0"

2 x 4 x 8' long

Nail one of the short 2-x-4s between the two 8-foot (or whatever length you use for your cabinets) lengths at each end. Nail in at least two other lengths, dividing the space evenly between the two ends of the frame.

You can now construct another framework just like the first, except that the top members of the framing can be 2-x-2s rather than 2-x-4s. When this is complete, you will have a second frame the same size as the first.

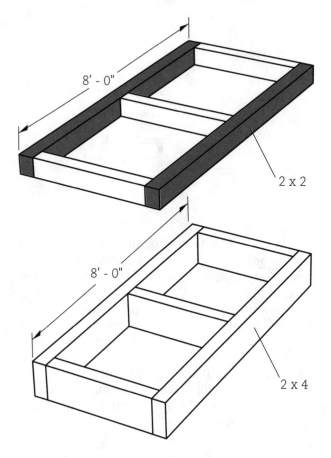

8' - 0"

2 x 2

8' - 0"

2 x 4

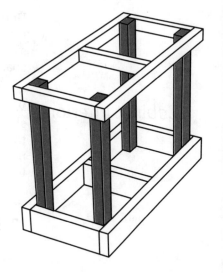

Cut additional 2-x-2 lengths to fit at the corners of the two frames. You can install them inside the four corners if you wish. They should be 3 feet long, and they should be nailed in place in a true vertical position.

Measure, mark, and cut the end units. These should be 3 feet high and 27 inches long (the extra 3 inches to allow toe space under the cabinets). The end pieces should be cut out on the lower outside corner so that the toe space will be 4 inches high and 3 inches wide.

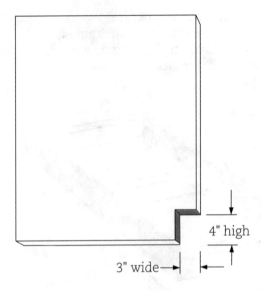

4" high

3" wide

Nail the end pieces in place on both ends of the framework. If one end is to be positioned against a wall, you can use less expensive panels for this work. You can also use thinner and less expensive panels for the backs of the cabinets. Once the ends are in place, measure, mark, and cut the bottom pieces. These should be just the right size to fit against the end pieces, but the corners will need to be notched so that they will fit around the corner posts.

Upright lengths should be installed wherever doors will be mounted and wherever doors will meet or close on the front side of the cabinet. On the back side, install matching uprights spaced exactly as the front ones are.

Cabinet interiors

When the uprights are in place, determine the height of shelves, if any, and cut 2-x-2 lengths to run from front to back to match the locations of the uprights, including the corner posts. You can toenail these support units or use corner braces to hold them in place.

You can now cut and install the shelf for the base units. Mark on the wall and floor where the unit will be placed. Leave enough space for the *carousel unit* to be installed in corners.

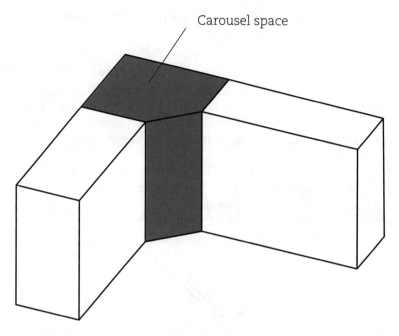

Carousel space

Frame and complete the other units that will form the right-angle cabinets for the room. Construct them as you did the first unit. When the second unit is complete to the same point as the first one, position it along the wall in its permanent place.

The front edge of the two end pieces should not be closer than 1 foot to 15 inches. The space in the corner is reserved for the carousel unit.

Measure diagonally across the floor from one front edge to the other and from the front edges to the back edges. Cut a framing member to reach diagonally across the space where the two front edges are located. Complete framing for the carousel unit as you did the other units. The major difference is that the framing will have five sides rather than four. The fifth side is the short diagonal running between the two front edges. Measure and cut out the floor or bottom of the carousel unit. Nail it in place atop the framing members.

When all base units are complete to this stage, measure and cut tops from ordinary plywood stock. On the front, cut out the stock for the drawers. Mark and cut out rectangles slightly larger than the drawers will be.

Building drawer guides

From the facing between drawers run a drawer guide from the facing to the back of the framing. This guide is installed so that it will be at the height of the bottom of the drawer. If you need to, install a length of 2-x-2 or similar stock along the back of the framing to hold the back end of the drawer guide.

You can make a drawer guide by nailing a 1-x-2 length on top of a 1-x-3 length of similar stock. Center the 1-x-2 so that a right angle is formed by the sides of the two units. Do this on both sides of the drawer, and the drawer will be guided between the two guides.

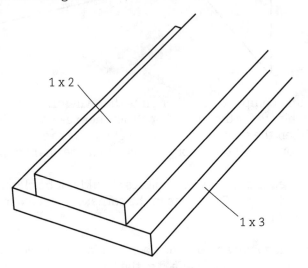

1 x 2

1 x 3

The left side of the assembly will be the right guide for the first drawer and will also serve as the left guide for the drawer to the right of the first one. Check to see that the top unit aligns with the two drawers. You will need a guide on each side of all drawers.

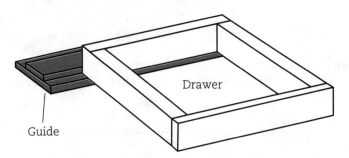

Guide

Drawer

When the framing is complete, cut and install the finish facing for the cabinets. This is where expensive wood and careful cutting come into the picture.

When you are cutting this facing, if you are using a circular saw, remember to cut while the facing stock is facedown. The splintering will then be on the back side rather than the front.

Sand the edges until they are perfectly smooth. Attach by using tiny color-matched nails or by using wood glue. The glue leaves a very smooth and uninterrupted surface.

When you attach doors, one common method is to cut the door 2 inches wider and taller than the door opening. Use offset hinges on the doors, and when they are closed, the outside edge and the top and bottom edges will lap the door opening rather than fit inside the door space.

Installing finish facings

Door and drawer assemblies

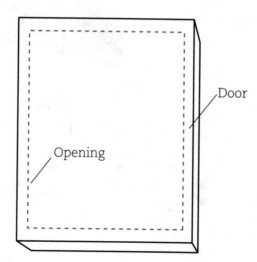

Door

Opening

You can install roller assemblies for smoother and easier operation. If you do not wish to go to this trouble and expense, you can construct a simple drawer and slide it along the guide surfaces.

To make the simplest type of drawer, do not buy special equipment for rabbet cuts or dadoes. Use ordinary materials and assemble them in the following manner. Use a board 4 or 5 inches wide (depending upon the size of drawer) and ¾ to 1 inch thick. Cut the board the proper length so that it is at least 1 inch and not more than 2 inches wider than the drawer opening.

Cut similar boards for the sides of the drawer. These boards should be no thicker than ½ inch. Attach these to the outside board by using small finishing nails or glue. You can also use very small bracket braces installed inside the drawer.

Cut the end piece and attach it to the ends of the side pieces. Let the end piece lap the entire thickness of the side pieces. Use finishing nails or wood glue, or both, to fasten the end pieces.

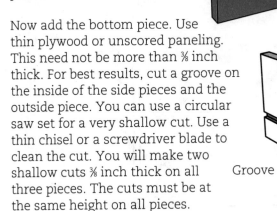

Now add the bottom piece. Use thin plywood or unscored paneling. This need not be more than ⅜ inch thick. For best results, cut a groove on the inside of the side pieces and the outside piece. You can use a circular saw set for a very shallow cut. Use a thin chisel or a screwdriver blade to clean the cut. You will make two shallow cuts ⅜ inch thick on all three pieces. The cuts must be at the same height on all pieces.

Groove

When the bottom piece is cut, insert it into the grooves before the end piece is attached. Then let the end piece set on top of the bottom section. Add a drawer pull to the assembly, and the drawer is complete.

Construct the wall units for the cabinets in the same manner. The difference is that you will not use stock larger than 2-x-2; 1-x-2 stock will usually work well.

When the units are complete, you can add the facing now or wait until the units are hung. It is easier to do the final work while the units are off the wall and in a more comfortable work position.

Now you are ready to hang the cabinets. This is reputed to be very difficult, but it is really quite simple and easy to handle.

Whether you make them or buy them, you can hang them and save some money. Some cabinetmakers will not install cabinets on log walls or other uneven surfaces. They also point out that if studding is not really solid, the nails or screws might not hold, and the cabinets could fall.

If you have used good studding, you should have no difficulty securing the cabinets. Your major problem is to lift and hold the cabinets at the proper height if the cabinets are mounted at eye level and above.

Install shelves by attaching a small support piece on each side of the cabinet. This can be as small as ½ inch by 1 inch, with the broad side attached flush against the side of the cabinet. The shelves then sit on the support pieces on each side of the cabinet. Use tiny nails or glue to hold the shelves in place.

When the cabinets are delivered or complete, you should mount the higher ones first. You do not want to work over the lower cabinets and risk damaging them, nor do you want to experience the difficulty of trying to work at arm's length.

There are two simple ways of mounting the cabinets to the wall. First determine the proper height. You must mount the cabinets at least 24 inches above the range burners for safety. Mount the cabinets low enough to reach them without difficulty and high enough to look attractive.

Constructing and hanging wall units

To mount the cabinets, you will need a power drill and some screwdriver attachments. You will also need 3-inch screws with flat heads.

Measuring for cabinet placement

Set up supports to hold the cabinets in place when you lift them. Do not try to hold the cabinets and drill at the same time. Set the cabinets on a foundation composed of stacked building blocks with boards across them or something similar. With the cabinets in place, drill a hole through the back and into the studding behind the cabinets.

You may have to measure to find the studding. The first hole should be three-fourths of the way to the top of the cabinet. When it is made, move to the same height and find the next stud.

Use a level to be certain that the cabinet is sitting in the proper position. Do not worry much about the vertical position of the front at this point.

When the holes are all made, use the power drill with a screwdriver blade in the chuck to drive the screws quickly. Sink the screws until they are almost completely driven. Then use a level to determine the exact reading up and down the front of the cabinets. When you are satisfied, sink the screws completely until they pull the cabinet unit tightly against the wall. Now remove the support system and use the level once again to double-check the position of the cabinet.

Install the remainder of the screws, screws in every stud no more than 8 inches apart up and down the studding.

When all wall units are installed, move the base units into position. Use a level to get the proper reading from front to back and from side to side. If you must, use shims under the framing to get level readings. Use screws again to attach the units to the studding. Drill holes into the studs, then use screws that as above, are slightly larger than the pilot holes drilled.

The second method of installation is to lay the units on a piece of plywood, then mark around the outside edges of the cabinet so that when you cut along the lines, the plywood is exactly the same size as the unit. You can cut inside the lines slightly to make the plywood a little smaller than the cabinet.

Mount the plywood on the wall by using either 16-penny nails or 3-inch screws, or both. Use the level to determine the correct position of the plywood. When the plywood is mounted, raise the cabinets as you did before. Align the cabinets with the plywood, then attach the cabinets to the plywood. If you do it in this fashion, you do not have to worry about finding studs. The plywood is securely mounted, and if you attach the cabinets to the plywood securely, your work is done properly.

All that remains is cutting out for the sink and attaching work surfaces. To do the sink marking easily, turn the sink upside down and mark around it once the sink has been positioned properly. Then cut ½ inch inside the markings so that the lip of the sink will hang from the plywood.

Drill holes for the sink connection. You can use a circular-saw attachment that can be installed in your power drill.

To install the laminated cover of work surfaces or counters, lay the cover in position and mark the underside where it hangs over the edge of the cabinets. Turn the laminated cover over and cut along the mark. This material is very hard to cut, and you may need to use a jigsaw or hacksaw.

When the cuts are made, place the covering back in position to see that you still have a correct fit. You can extend it over the outside edge slightly, no more than ⅛ inch.

Use a brush to cover the top surface of the plywood and the bottom surface of the covering with adhesive. When you start to install the laminate, it will adhere instantly to the plywood, so you cannot place it and then try to adjust. You must position it exactly the first time.

Start by lowering the back part of the covering to the plywood. Hold the covering so that it will not sag and touch anywhere else. Lower slowly from both ends. When proper contact is made, slowly lower the front part of the covering until it is positioned flat across the entire surface.

Later you can cut thin strips to fit under the slight lip you left extending past the edge of the plywood. The strip will create an even corner. Install it exactly as you did the top covering.

Making doors

WHEN YOU ENTER the final stages of finish carpentry you will find that some of your greatest expenses await you. You were able to install all the floor joists in the house for less than it will cost to install windows and doors in a single room. Three windows will cost as much as all your subflooring.

Although windows and molding are extremely expensive items (some molding running over $1 per foot for a thin sliver of wood and some window units costing well over $500), doors are somewhat less expensive if you buy veneer hollow-core doors, which work reasonably well for your interior needs.

If you want a truly sturdy door that combines rugged good looks with endurance, strength, and simplicity of design with very low cost, you can make your own very simple doors. You can also make your own shutters by using the same basic approach (chapter 14).

The doors described here are best suited for the basic house with leanings toward Early American.

Tongue-and-groove timber doors

For an extremely strong and attractive door, use 2-x-6 tongue-and-groove timbers. You may have had some left from your subflooring work or roof sheathing if you used the tongue-and-groove timbers that serve as sheathing and ceilings.

Determine length and cut 2-x-6s accordingly. A good height for a door (and this depends largely upon the rough door opening you left earlier) is 6 feet and 8 inches, or 80 inches.

If you have 12-foot timbers, you can use one full-length piece alternated with a 4-foot length and one that is 6 feet 8 inches. Use timbers with good clean appearance on both sides.

For a simple Early-American design, lay the 2-x-6s on a work surface, match the tongues and grooves, and pull the units tightly together. Across the top and bottom install a length of the same stock (with tongue and groove cut away with a circular saw) and nail it to each unit of the door.

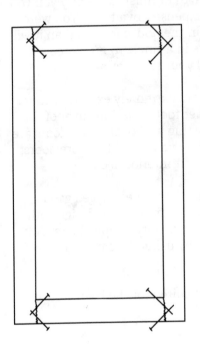

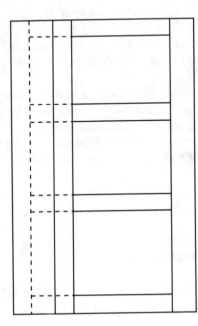

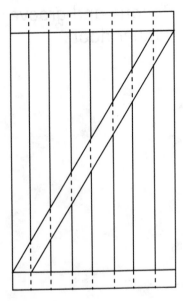

When top and bottom lengths are in place, cut a third piece to make the traditional Z pattern. Nail the piece in place and use a punch to sink all nail heads until they are even with the wood surface or slightly below it.

When the door is complete to this point, drill holes in each of the units and through the top and bottom pieces as well as the diagonal length. Use a ¼-inch drill, then use ¼-inch bolts, nuts, and washers.

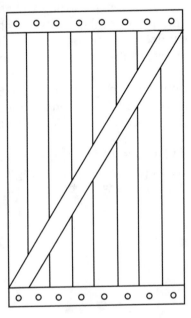

Before installing the nuts and bolts, on the back side of the door use a paddle-blade drill to make a hole just deep enough to allow you to sink the nut until it is even with the surface of the wood. Then install the nuts and bolts, tightening the nuts securely.

When this is done, use a hacksaw to saw the excess length of the bolts off so that you are left with a smooth finish. When you are certain you do not want to modify the door, you can use the peen of a hammer and tap the ends of the bolts to flatten the bolt end even more. At the same time, you render the nut almost impossible to remove.

When you hang the door, install it so that the round heads of the bolts are on the outside, the nuts on the inside. This type of installation offers greater protection against intruders.

A modification of the basic door uses threaded bolts through the entire door. To drill for the threaded bolts, stand a 2-x-6 on edge and start the drill into the groove side, holding the drill as perfectly vertical as you can. Watch carefully to see that you do not let the bit run out the side of the timber.

Using threaded-bolt assemblies

When the hole is drilled completely through the 2-x-6, lay the timber on its side and fit a second 2-x-6 into place, with the tongue and groove aligned. Run the drill through the first timber again and into the second one an inch or two.

Remove the drill and the first timber, then continue to drill the hole the remainder of the way through the timber. When you are through, align another board and repeat the process of running the drill into the next board or timber an inch or two. In this manner you keep the drill holes aligned so that the threaded rod can be inserted easily. If you wish, you can drill through one timber and start the second, then insert the threaded rod through the first. When the second is complete, and the third is started, insert the rod farther. Keep up this pattern until the rod extends completely through the door.

When you tap the rod, do not hit the metal with a hammer. Always lay a piece of scrapwood over the rod, hit the wood, and let the wood drive the rod through.

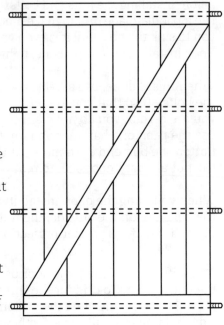

It is a good idea to use a paddle bit and enlarge the entry hole for the rod on one side of the door and the exit hole on the other side so that nuts can be countersunk. Before you attach the nuts, be certain that the door will fit properly. Once the threaded rods are in place, it is impossible to trim the door more than a fraction of an inch without removing the rods.

Insert a rod 4 inches to 1 foot from the top, an equal distance from the bottom, and one in the middle. You can place two in the middle portion of the door for greater strength and stability.

Trim away the tongue and groove on the outside edges of the door before completing the installation of the nuts and threaded rods. Countersink the nuts and use wood putty to cover the nuts if you prefer to keep them unseen.

Construction with 1-inch boards

If you have any 1-x-3 or 1-x-4 tongue and groove boards left from another job, instead of returning these for credit, you might prefer to use them to make another type of door. You can make a very attractive and neat door that is strong and useful by using the 1-inch or even thinner boards.

Start by using 1-x-3 or 1-x-4 square-edged boards laid in a rectangle the size of your door opening minus 1 to 2 inches. Fasten the boards together by toenailing, by using wood glue, or by a combination of the two. Lay the long boards in place and space them so that there is a distance between the boards that will align the outside edges with the size of the door opening. In other words, if your rough opening is 36 inches wide and you are using 1-x-4 boards (which are actually 3½ inches wide), you will need the boards spaced 29 inches apart, measuring from the inside edge of each board.

Cut top lengths to fit between the boards. At the first corner, lay the boards so that the edges are aligned correctly. Use finishing nails to toenail from the top edge of the short board into the inside edge of the long board. Then nail through the bottom edge of the short board and into the inside edge of the long board. For a stronger joint, coat the end of the short board with glue. Nail the board as before, and when the glue sets, the corner will be very strong. You can use 1-inch corner braces in each of the inside corners for even greater holding power.

Fasten all four corners as directed. Be sure that your total dimensions of the door will permit you to fit the door into the rough opening. If the door is slightly too wide, use a circular saw to rip the door lengthwise for a proper fit. You can let the boards extend past the bottom crosspiece for an inch or two until you are certain of the length needed. You can saw off the bottom ends without problems.

You can install one or two short boards across the middle portion of the door for greater stability. Keep the ends of the boards flush with the surface of the long boards.

When the rectangle is complete, start to install tongue-and-groove boards. Trim the groove off the first board and nail it in place (using finishing nails) flush with the outside edge of the long board. Use only two small finishing nails at the top and two more at the bottom to hold the board in place. You will strengthen the holding power of the board and door in a short time.

Install the second board by positioning it so that the groove will slip easily over the tongue in the first board. When the board is seated well, drive nails through the tongue and into the crossboards. Use at least two nails at every crossboard location.

Continue the installation across the door. If you wish to use shorter lengths of tongue-and-groove boards, let them end on one of the crossboards. This practice is not suggested but can be used if you prefer to make some use of board lengths that would otherwise be wasted.

When the side is complete, turn the assembly over and repeat the process on the opposite side. When you are finished, you will have a door that is two boards thick, plus the thickness of the boards forming the rectangle.

As a finishing touch, drill holes at each point where the long boards cross the short ones and install short nuts and bolts. Do not countersink the nuts. Instead of countersinking, tighten them until the nuts bite into the wood and sink themselves halfway into the wood. Use a hacksaw to cut away excess bolt and use a hammer to flatten the ends of the bolts. You can also use a file to smooth the ends of the bolts.

Such doors may sound complicated to construct, but in reality they can be built very quickly and with little cost. These doors would cost a great deal of money if they were available at supply outlets. You have the added satisfaction of knowing that you made the door yourself, saved money, produced a strong and attractive door, and created an original, unlike any door on the market.

Building doors with straight-edged boards

If you prefer to use straight-edged boards, you can construct the door as described above. The major difference is that when the door is complete, cracks will be visible. You can handle the problem in one of two ways.

First, install a layer of building paper over the rectangle when you construct it. The building paper will prevent unsightly cracks from showing. The paper will also help to prevent loss of cooling or heating.

The second method is to start on one side with full-width boards and on the reverse side use a board that is half as wide as the others or two-thirds as wide. Complete the side with a second thinner board. By doing so, you alter or stagger the board alignment and thus keep the cracks from being aligned. You can also use building paper between layers if you choose to use the staggered method of construction. Either way will produce an attractive and sturdy door.

There are variations of these basic doors that work effectively. One method is to secure a number of 2-x-4s as close to perfect on the edges as you can find them and lay one 2-x-4 on a work surface. Coat the top surface with glue, then position a second 2-x-4 on top of the first one so that all edges are aligned.

Lay three or four 2-x-4s together, one on top of the other, the edges aligned perfectly. Hold the timbers together with a pair of C-clamps. Mark points in the center of the top 2-x-4 4 inches from the top, 4 inches from the bottom, then divide the distance between the two points and mark two more locations so that you can drill holes through all the stacked timbers and have the holes spaced for maximum holding power.

When you have completed drilling, use one of the 2-x-4s as the top unit in a second stack and use the holes as pilot holes for the second drilling. For the third stack, use one of the 2-x-4s as before and continue this practice until you have four holes aligned as correctly as possible through all of the 2-x-4s to be used in the door.

When you are ready to begin assembly of the door, lay one 2-x-4 (with as close to a perfect face as you can find) with the best surface facing downward. Lay another 2-x-4 on top of the first one. Align the holes. It does not matter if the face of the 2-x-4 is defective. This surface will not show.

Use finishing nails or nails as large as 10-penny to 16-penny to hold the 2-x-4s together. The nail heads will not show, but drive the heads until they are perfectly flush with the surface of the wood. The glue will close the space between the wood and prevent air passage.

For an even tighter fit, you can use a pair of C-clamps to pull the first three or so pieces together. The clamps are not wide enough for any more than the first few units.

Variations of basic door construction

You may find that your greatest difficulty is in keeping the 2-x-4s rising in true vertical. You may need to lay the assembly flat against a smooth work surface to keep the sides of the assembly aligned.

Leveling door assembly

You can also use a level, just as you would on a block or brick wall. Check the level reading of each 2-x-4 as it is added to the assembly.

Make the level readings before you drive the first nail. You can drive two or three nails partway in, then check the level. If you need to make modifications you can remove the nails easily.

As you work, insert threaded rods through the holes. When the door is complete, the threaded rods will reach through the complete assembly.

When the door is made, you have a structure that is 3½ inches thick. The final touch is to tap the threaded rod back into the end 2-x-4 slightly, then use a paddle bit to enlarge the hole so that a nut can be countersunk. When the hole is ready, tap the threaded rod from the other end and force the end out far enough so that you can start the nut and tighten it until the end of the rod is flush with the outside edge of the door and the outside edge of the nut.

At the other end, tap the rod again and recess the end where you are working. Make the hole for countersinking the nut. Tap the other end once again until the nut and rod end are again flush with the wood. Then install the nut on the opposite end and tighten until the nut is flush with the wood surface. Saw off the excess rod and file or use a hammer to tap the end of the rod until you have a flush surface.

Such a door is expensive. You will need 24 2-x-4s, and each of these will cost more than $1 each. The rods will cost a total of about $12 at 1992 prices, and the combined total of all the materials will equal close to $50, again at 1992 prices.

To eliminate dating the costs, think in terms of equal value. For the cost of the door, you could buy three panels of medium-grade plywood, two and a half stereo tapes or two compact-disc recordings. You could also buy two tankfuls of gas for a medium-size automobile.

While the price may be slightly higher than those of most suggestions in this book, remember that the finished product is a fantastic door that will be unlike anything you can find on the market.

Remember also that here is a place to use marred or defaced 2-x-4s that you could not use elsewhere. As long as the edges are in good shape, the faces of the 2-x-4s are not important, except for the first and last ones. You can probably make a deal with a supplier to buy a number of rejected timbers that no one else will buy; by doing so, you can build the door for less than half the originally stated price.

You can combine several techniques mentioned in this chapter to create your own individualistic doors. These doors can be left in their natural color for special blending with wood wall coverings, or they can be stained and water-sealed and used as exterior doors.

CHAPTER 14

Building shutters

ONE STRONG interior shutter can be made in the same way you made doors. Use 2-x-6 tongue-and-groove timbers cut to fit inside the window framing.

Beginning steps

Your first step is to divide the distance across the window by 2. If the window is 36 inches wide, you need to build two 18-inch shutters for the window.

If your shutter stock is a true 6-inch timber, three lengths of lumber will build each side. You need to allow a fraction of an inch, ¼ or so, on each side of the window for hinge space, unless you decide to recess the hinges.

Cut two lengths of 18-inch stock to fit across the top and bottom of each shutter. Measure the window from top to bottom for the proper length. Cut three sections the proper length and lay these on a work surface. Lay two crosspieces so that one will be located 2 inches from the top and the bottom of the shutter. Trim the groove side off one of the shutter lengths, then lay the length on the crosspieces. Check to see that all the alignments are correct.

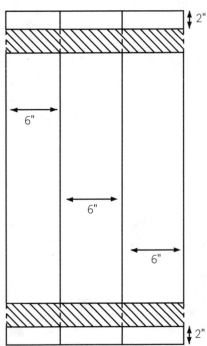

Use very small finishing nails to hold the shutter section in place. Do not drive the nails in all the way; leave at least ½ inch sticking out. You will later use nuts and bolts, but at this point you can let the finishing nails hold the section in place until you are certain that the fit you want is achieved.

Fit the second length or section so that the tongue and groove fit together well. Use cut nails to nail the second section to the crosspieces at the top and bottom. You can use finishing nails if you prefer.

When you are ready for the third section, use a circular saw to cut off the tongue section. Use small finishing nails to hold the section in place.

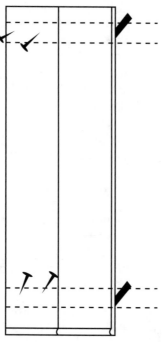

When the section is complete to this point, use a ¼-inch bit to drill two holes in the shutter sections at the top and two more at the bottom. Be sure to pull out one nail before you start to drill. Start the drill in the hole made by the nail so you won't have unnecessary marks scarring the wood. Drill the hole slowly. As soon as you feel the tip of the bit start through the wood on the back side, stop and turn the section over. Remember that the wood will be splintered slightly where the bit emerges. Insert the tip of the bit into the hole where the bit started to emerge and drill back in the opposite direction.

Drilling bolt holes

You will now have a clean hole through both the shutter length and the crosspiece. Move the bit back and forth while you run the drill at a moderate speed to clean out any shavings or other residue.

Use bolts that will reach through the wood and finish flush with the surface of the wood. Before you install the bolts, use a slightly larger bit and drill a shallow hole large enough for the

Installing bolts

nut to fit inside it. Your bolts should be smooth-headed (*carriage bolts*) so they will fit tightly against the shutter.

Run the bolt in from the back side so that the smooth head shows from the outside. If you should have an attempted break-in, the prowler will not be able to use a wrench or other similar tool to loosen the bolts. On the inside, install the nuts and use a socket wrench to tighten the nuts all the way until the two sections are pulled snugly together. Do not tighten completely at this point.

Install all of the other nuts and bolts into the shutter section. When it is complete, you can tighten the nuts until all of the sections and crosspieces are pulled tightly together.

Complete the other section of the shutter just as you did the first. When it is complete, you can install the two sections.

Setting hinges

To recess the hinges, set the hinges in place and mark around the outside with a sharp-pointed pencil. Use a chisel and hammer to cut out the recess portion. Set the chisel so that the beveled side is facing the inside of the marked rectangle. With the chisel blade resting precisely on the marked line, strike the handle of the chisel and drive the blade into the wood ⅛ inch.

Move the blade so that it overlaps the first cut slightly and make another cut. Continue doing this until the entire hinge recess has been outlined. Turn the chisel so that it is at a 45-degree angle to the wood. The beveled side of the blade should be facing the wood. Set the blade an inch away from the cut mark on either end of the rectangle and tap the handle lightly with a hammer. You do not want a deep cut.

When the wood starts to curl up in front of the chisel blade, you can lower the angle even more and tap very lightly until the blade tip reaches the cut mark and the chip or sliver of wood pops free.

Repeat this process at the other end of the rectangle. Then cut out all of the higher wood surface between the two ends of the rectangle. The final cuts should be parallel to the long cut line along the sides of the rectangle.

When the cutout is complete, set the hinge inside to see if the fit is good. The top surface of the hinge should be perfectly flush with the surface of the wood.

Mounting shutters

Place the other section of the hinge against the window frame, mark as you did before, and make the cutout. When you are finished, the two hinges should seat precisely and the shutter section should fit neatly inside the window frame.

The final step before mounting the shutter section is to lay the section on a work surface so that the crosspieces are facing upward. Mark a line across the end of each crosspiece 1 inch from the end opposite the hinges.

Use your circular saw set on bevel cut and make the cut across the end of the crosspiece. The purpose of this cut is to allow the shutter sections to close easily. If you don't make this cut, the ends of the crosspieces will meet and not let the shutters close completely.

Installing locks

When both shutter sections are installed, you can install a locking device in one of several ways. You can use a slide bolt installed across the junctions of the two sections. If you install the bolt assembly so that the base metal extends ⅛ inch or slightly more, the window will be even more secure. Even a hacksaw would encounter difficulty in cutting through the base metal and bolt.

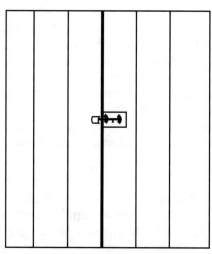

A second way is to install the slide-bolt assembly so that the bolt is pointing straight down from the bottom of the shutter. When you want to lock the

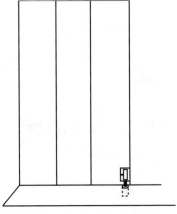

shutter, you slide the bolt downward into a hole drilled in the windowsill.

You can also install the bolt assembly so that it points upward, placing it at the top of the shutter. When you lock the shutter, the bolt will slide upward into the top of the window frame.

You do not have to use 2-x-6 shutter material. You can use 1-inch tongue-and-groove boards for a sturdy shutter. Install the hinges across the edge of the shutter where the crosspieces are located, and you will have abundant room for the hinge to be seated fully.

Advantages of shutters

Consider the economy of the shutters for a moment. If you build neat and well-fitting shutters, you don't need to buy window shades, curtains, or drapes. The money you save on the drapes or curtains will offset a huge part of the cost of the shutters.

If the shutters fit well and you use molding inside the window frame or allow the window stile to act as molding, you also insulate the window areas well. You know that a huge part of your loss of cooling and heating occurs at windows and doors. You will save considerably on heating and cooling expenses.

You will also find that wood shutters will block out excessive sunshine, and that will not only reduce the load on the air-conditioning system but also prevent the bright light from bleaching or fading the fabrics of curtains, carpets, and furniture.

When you open the shutters, the two sections will fold back against the wall. You can use one huge shutter if you prefer, but you will need a great deal of room to swing the shutter open, and this may not be practical.

Mounting shutters in concrete walls

Such shutters work exceptionally well in basement windows that are set into concrete-block walls. To install these shutters into concrete- or cement-block walls, you will need to drill small holes (the size depending upon the size of the screws and anchors you plan to use) into the blocks, then use long screws through a panel of wood (1-x-4) and into anchors that you drive into the drilled holes.

To determine the exact locations for the anchors, set the board into the window space and hold it in position. Using a drill and masonry bit, drill a hole through the wood and into the cement blocks just enough to mark the drill locations.

Take the board down and continue drilling where the marks are located. When the holes are drilled, insert the anchors by pushing or tapping them gently into position.

A second method is to use steel or masonry nails and simply nail the board to the inside of the window frame. Be careful when nailing steel nails. The heads are very prone to break and fly across the area with sufficient force to cause serious injury to eyes. It is always advisable to wear protective eye covering when hammering masonry nails.

You can install the shutters on the outside of the windows if you wish, but this is rather impractical. If you want to use the old-fashioned shutters that close from the inside, you can use sections made as suggested above and install them with the hinges on the outside.

Hooks and latches can be installed to hold the shutters flat against the exterior walls when the shutters are open. The same latches can be used to keep the shutters closed.

Shutters used only for cosmetic purposes can be made by using two or three narrow board lengths fastened across the crosspieces. You do not need to use hinges if the shutters will not be functional. You can use nails or screws to fasten the shutters in place. You do not need to have the boards touching. They tend to look better if they are spaced with 2 inches or so between them.

You can use many of the excess lengths of wood left over from longer timbers that had to be cut. This is a good way to make use of materials that otherwise would have been wasted.

CHAPTER 15

Installing interior molding

INSTALLING MOLDING is one of the final parts of finish carpentry, and on the surface it appears to be one of the easiest aspects of house construction. The job can in fact be very easy and fast if you will observe a few basic principles of installation.

You may wish to start work by installing *baseboard molding*. This material is usually thin (½ inch or similar thickness), 3 to 5 inches high (4 inches is a popular height), and manufactured in a variety of lengths. You will want to keep jointing at a minimum, so buy the longest stock you can find if you plan to use it in larger-than-average rooms.

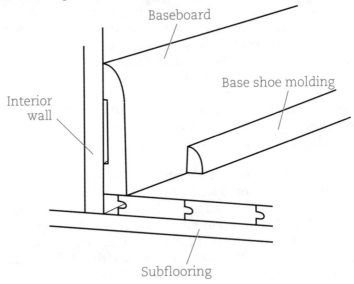

Baseboard

Base shoe molding

Interior wall

Subflooring

Planning installation

The disadvantage to buying longer stock is that you must be very careful in transporting the stock. If you buy 12-foot molding, it is a good idea to tie or tape several pieces together for hauling.

If five or six units of baseboard molding are taped tightly together, the entire package assumes a greater rigidity than one piece alone would have, and you are less likely to break or damage stock. You can tape a dozen or so pieces of quarter-round molding or cove molding together for hauling.

When you are ready to install, plan your work so that you will have the least waste and the smallest number of joints. Remember that baseboard molding has very few uses other than its primary purpose, and you need to keep room dimensions in mind whenever you are buying the stock.

Assume that you have a room 16 feet long and 15 feet wide and that the stock you need is available in 8-, 10-, and 12-foot lengths. Two 8-foot lengths will cover the longer walls, and you will have no waste. For the shorter walls, one 12-foot length will be 3 feet short, and if you cut a 3-foot section from another 12-foot section, you will have 9 feet left.

If you have other rooms with similar lengths or widths, you can use the 9-foot section by cutting it into three 3-foot sections. If you have no other rooms in which to use the leftovers, you can use two more 8-foot sections and have only 1 foot of waste.

To install baseboard molding in a corner, get an accurate and neat fit by placing one length so that it fits neatly and snugly into the corner and holding it there while you push the end of the meeting piece (after it has been miter-cut) against the first piece. Now use a pencil to mark along the side of the outside piece and on the inside piece. To put it another way, miter-cut the end of the piece that will be the outside section. The cut is perpendicular and slanted to the back of the piece. When the mitered end is pushed into the corner, hold it in place until the outside piece is pushed against it, then make the mark.

Take the pieces out of position and use a coping saw to cut along the curved line of the mitered end of the cut piece. When you are ready to install, place the uncut piece into position and use finishing nails to attach it to the studding of the wall frame.

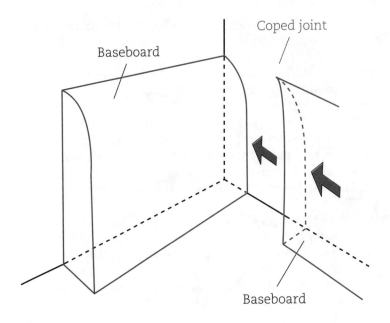

Baseboard

Coped joint

Baseboard

Then place the mitered and coped end of the cut piece and fit it against the segment that has been installed. Nail it to the studding of the wall frame.

Marking for installation

Before you begin installation, you may want to mark the wall or floor with small pieces of masking tape to show the location of the studs. The baseboard molding is typically not installed until the wall covering, including paint, and floor covering, including carpet, have been installed.

When you are joining straight-run baseboard molding along the length of a wall, use a miter cut to make the fit so that it will be less conspicuous. Rather than butt-join to straight ends, miter both ends, so that the two cuts are in the same direction, and join them so that the two cuts fit together smoothly. Such joints are very attractive.

You can also think of how the room will be used after the molding has been installed. If a long couch is to be used in a centered position along the wall, be sure that the joint occurs behind the couch.

For outside corners, miter both pieces so the ends will fit together neatly. Install the first section and let the mitered end extend to the point of the start of the cut. That is, let the beginning of the cut (on the back side of the baseboard) rest so that it is flush with the corner point. The long end of the cut will extend past the corner slightly.

When the second section is to be installed, it will be positioned exactly the same way. In this manner the two short ends of the cuts will both be flush with the corner point. The two long ends will extend exactly the same distance past the corner and will therefore meet flawlessly if you made good cuts.

To make miter cuts, you can use a miter box or a circular saw set at the proper angle. The miter box tends to produce more accurate and neater cuts. If the circular saw is allowed to veer even slightly from the cut line, the results can be unsightly.

Miter cutting

When you use a circular saw, remember that the teeth cut upward; therefore, you should cut with the back side of the stock turned upward. Otherwise, you will splinter the face of the wood.

If you do not have a miter box, you can construct a good one in minutes from some 1-x-4 stock. You will need three pieces of stock about 2 feet long. The box can be longer or shorter, depending upon your needs and the lengths of wood you have available.

Lay one board flat on the work surface, then stand two boards upright so that they are against the sides of the flat board. This is the position in which you will want to nail the boards. Use 8-penny nails to hold the sides in place.

Lay the framing square across the open end of the assembly so that the outside edge of the blade is aligned with the outside edge of the upright board and the tongue extends across both of the upright boards. Mark along the outside edge of the tongue. The mark should cross both boards.

Measure the total distance across the assembly, including the two upright boards, to the outside edge. Mark off the same distance from the line you just made down the top of the assembly. Use the square again to make a similar mark, so that you have outlined a perfect square atop the upright boards.

Use the square to mark a diagonal cut line from the outside corner of one upright board to the outside corner of the other upright board. Do this in both directions so that you have made a huge X across the square.

Using a circular saw or handsaw, and cut along the diagonal lines in both directions. The cut should extend all the way through the upright boards down to the surface of the flat board. You now have a cut line for both types of angles needed.

Move 6 inches or so from the diagonal cut lines area and mark a cut line straight across the assembly. Use the saw to cut along the line as you did before.

The miter box is now complete. You can use it for straight cuts and diagonal or mitered cuts. Depending upon the type of cut you need, you can lay the stock flat or stand it up to make the cut.

Installation in corners and alcoves

In and other small areas where there are angles and corners, measure the length of cut needed and make the miter cuts or cope cuts as needed. Where the baseboard molding meets door framing, use straight cuts.

You may wish to use shoe molding at the bottom of the baseboard, as shown earlier. This is optional: some do-it-yourselfers like the look of added neatness and trim provided. Use it if the baseboard molding does not conform to the floor as accurately as you prefer or if you need wider molding to cover the edge of the carpet or other floor covering.

To install shoe molding, cut the miter joints exactly as you did for the baseboard molding. Use the same methods to cut for inside corners and outside corners.

Around ceiling-wall junctures you can use cove molding or bed molding. These are shaped so that they will fill the angle formed by wall and ceiling and extend in both directions to cover any problems in wall or ceiling coverings.

To cut corner fits, tack one segment of molding in place by pushing it into the corner snugly and then using two small finishing nails to hold the molding in place. As before, push the second section against the first one and make a pencil mark around the first segment, the one that extends into the corner. Cut along the line. Nail in the square-ended molding, then push the cut molding into place and nail it. Be sure to nail into studding or ceiling joists.

Crown molding is highly expensive, but it is a popular type of finishing touch added to many rooms. Here you will use the same methods of installation, but you must measure carefully and cut with even more care. When you ruin a length of crown molding, you have wasted a great deal of money.

Some builders prefer to cut the mitered end first and then cut the miter-lap end. To do this, you need a length of molding that does not reach the full length or width of the room. Make the miter cut for the corner, then hold the section of molding in place to see that it will fit properly. Install it temporarily by using two or three finishing nails, just enough to hold it in place. Then make the miter cut for the other corner and hold the second length so that the two pieces overlap, so that you can see exactly where the two will meet.

Mark the point where the lengths will meet and make the miter-lap cut on one of them. Finish installing the piece that will allow the miter-lap cut to be exposed or face into the room. Hold the second length in place again to see that the marks conform as they should. Take down the second length and make the opposite miter-lap cut. When you install the length, the laps should meet and fit together exactly. You already know that the corners are correct, and you will be assured of a good fit across the entire wall area.

Other carpenters like to make the miter-lap cuts first and then make the corner miter cuts. They assume that if there is to be a visually bothersome problem, it should be in the corner where there is less visibility.

Completing finish work

Another thesis is that on corner cuts you can fill any small gaps with wood putty or similar material that can be sanded and painted when the substance is dry. If wood-puttying is done well, it is almost impossible to see the places where the corrections were made.

Our advice is to start installing molding in rooms that will be seen least. Choose workrooms, hobby rooms, or second bathrooms for your early work. Do the finish work in the basement first if you plan to have a basement that is used as part of the living space.

If you have problems, they will not be as intrusive in the decor of the house as they would if they were in a den, living room, or dining room. Learn from your mistakes in low-visibility locations and avoid these mistakes in more public areas.

Finishing doors and windows

You need to install molding around doors and windows as well. These sections of the room are finished in the same way as other moldings are installed. Typically you want to use the mitered-corner approach, so that every window and every door, after the traditional casing is installed, is finished in the same fashion.

Miter joint

Casing

Door frame

You can start with the side lengths and miter-cut them so that the short end of the cut on each end will align with the corner of the window framing. Some prefer to nail up the side length on one side of the window, then miter-cut and install the top or bottom lengths. By using this approach, they install three sides, then fit the fourth side into the space left. You can also install the two side lengths, then cut and fit the top and bottom lengths into position.

If you make the measurements and cuts so that the short end of each cut is aligned perfectly with the corners of the window framing, you will have very little difficulty. If you install the side lengths first, measure the distance between the two long points of the side lengths. Measure next the distance between the two short points between the two installed lengths. When you mark for the miter cuts, measure again to be certain that when cut, the length will fit perfectly into the remaining space.

Do not be overly concerned about your ability to install window-trim molding that will fit exactly. If the window framing is square and you make the cuts straight, the fit should be no problem as long as you make correct miter cuts.

When you make the cuts, hold the molding stock firmly against one side of the miter box. Do not let the molding slip or move as you cut. If you are confused about which type of cut to make, lay the stock in the miter box and make a light mark across it in the direction of the cut you plan to make. Hold the marked stock up to the window or other location to see that the cut line is accurate. If you see that your mark is wrong, take down the piece and re-mark it.

Take care to position the cut mark so that it aligns with the slots for making the miter cuts. If you are not careful, the cut may be too short for the intended use of the molding.

As you work, remember the old advice about measuring twice and cutting once. We have stressed this element because of the cost of molding and because molding is one of the items that cannot be used in many other spaces. If you ruin several lengths, you have negated much of the savings you anticipated by doing it yourself.

CHAPTER 16

Trimming doors & windows

DOORS AND WINDOWS have been focal points in the modern house for many decades. Visitors to a house are of necessity brought into close contact with doors in every room. Even in doorless entryways some form of trim is used to mark the point where the wall covering stops and the entry begins.

The trim is usually wide boards or special molding from 5 to 6 inches wide. The molding covers the entire perimeter of the door, except for the floor area. Window trim usually completely surrounds the entire window.

This trim is very easy to install. You will not encounter any difficult problems, and you will not need special tools or equipment.

On the interior side of all doors you need to install the door trim or framing and facing. The rough door and window openings have a framing inside them, but there is sometimes a gap between the inside framing and the wall framing. When wall coverings are installed, the coverings often stop at the end of the studding. The trim is needed to conceal the gap between the two materials.

Cutting trim materials

You can buy special door and window trim, but the least expensive way is to use 1-x-5 or 1-x-6 boards that are free of defects, straight, and sound. If a board has a defect on the back side, you can still use it as long as the face is flawless or nearly so.

Most do-it-yourselfers prefer to use square-cut framing around all windows and doors. A few prefer to make miter-cut trim fittings.

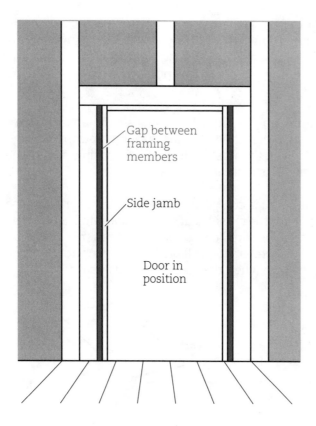

Gap between framing members

Side jamb

Door in position

An easy way to install the square-cut trim is to measure from the floor to the top of the door opening, then cut boards to fit on each side. When you make the cut marks, be sure that the blade of the square is set snugly against the side of the board so that you get a perfect right angle to the side of the board.

If the cut is even slightly less than accurate, there will be a gap between the end cut of the board and the top piece of trim that rests upon the end of the board you are cutting. This gap will be noticeable even if you use wood putty to make emergency repairs.

You can make a faster cut by using a circular saw; however, if you deviate from the cut line, you will mar the end of the board. A handsaw is much slower but more accurate; another plus is that the handsaw does not splinter wood as the circular saw does.

To minimize splintering when using a circular saw, turn the board over so that the back side is facing upward. Any splintering will then be on the back or invisible side of the board when it is installed.

When the board is cut, position it so that the door-side edge of it is aligned with the outside edge of the side jamb. Some builders like to recess the trim boards ⅛ inch, and you may install in this fashion if you prefer; you will prevent the door from scraping against the trim when the door is opened and closed.

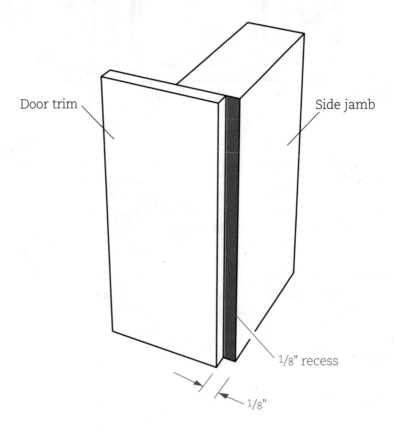

Door trim

Side jamb

1/8" recess

1/8"

While holding the trim board in place, drive finishing nails through the board and into the edge of the side jamb on one side of the board and the studding on the other side. Use finishing nails large enough to hold the trim securely but not large enough to detract from the appearance.

You can use a punch to set the nails, or recess them, after the trim is in place. It is a good idea to leave the nail heads sticking out an inch or so while you make the rest of the trim installation. You can set or recess the nails when the job is done and you are satisfied with the results.

Mark and cut the second board as you did the first. When it is cut, nail it in place and again leave an inch of nail sticking out in case you need to make corrections.

The final step is to mark and cut the top trim board. Measure from the outside edge of one side trim board to the other. The top trim board should reach completely across the tops of the side boards.

Final trim steps

Cut the top trim board and install it. Check to see if there are poor fits or gaps anywhere. When you are completely satisfied with the trim, you can seat the nails the remainder of the way, then use the punch to set the nails.

To set, recess, or countersink the nails, hold the punch so that the narrow or pointed end is set against the head of the nail. Hold the punch firmly in place, then tap the broad end of the punch with your hammer until the nail heads are slightly below the surface of the wood. You can use wood putty to fill in over the heads if you wish. The nails are then completely invisible. If you do not want to use putty, you can paint the boards later, and the paint will fill in the holes and conceal the nails.

On interior doors, trim both sides in the same manner as the door you just completed. You can later add molding if you wish, although this is a purely optional step.

Trim windows the same way. Align trim boards with the window framing and use finishing nails for installation. You can also use molding here.

Miter-cutting trim materials

If you want mitered trim work, measure the distance from the floor to the top of the doorframe and make a note of the distance. Then measure from the top of the frame to a point equal to the width of the top frame boards and add that distance to your side trim boards.

Cut the board according to the longest marks. Then make a miter cut from the mark denoting the top corner of the doorframe to the top corner of the top board.

Do this on both sides of the doorway. Install the boards and leave the nails protruding slightly so you can remove them if necessary.

Measure from the inside corner of the doorframe to the corresponding point on the other side of the door. Then measure the distance from the points of the outside edges of the side trim boards.

You should double-check all measurements and marks before you cut. Use the square or straightedge to mark the diagonal cut lines across the top trim board.

Fit the cut trim board into position. If the fit is good, nail it in place. If the fit is not satisfactory, remove the board and make the necessary adjustments.

You can also install the top trim board first if you wish. The diagonal cut lines should take into account the width of the side boards and should extend from the corner of the doorframe to the outside edge of the trim boards.

If all stock is the same width, you can make a diagonal cut by using the miter box. The fit should be perfect. The other measurements are necessary only if the widths of the boards are not the same.

Trim windows the same way if you prefer miter-cut trim work. Do not attempt to nail too near the point of each angle or diagonal cut.

If you find that your nails are splitting the wood, there is a quick and easy solution to the problem. Stand the nail on a solid surface, such as the top of a piece of thick metal or other very firm surface, and while you hold the nail still, strike the point of it with the hammer and flatten the point somewhat.

Preventing splitting

This may sound totally illogical, but it works. A sharp point will split wood readily; a blunt point will not. This is a trick that cabinetmakers and other craftsmen who work with thin and soft woods learned long ago. Even people building log cabins learned that if they used a spike with a flat rather than pointed end, they could drive it through a log and into another one and not split either log, while a pointed spike might damage either or both logs.

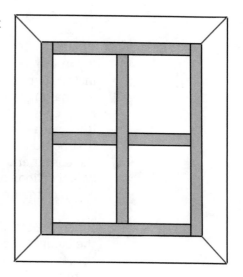

When the trim is complete, if you want to add additional trim or decorative molding, you can do so easily and quickly. Miter-cut the molding just as you did the trim boards. Measure and mark the same way.

Most people install the baseboard molding and butt it against the door trim boards before any decorative molding is used. The decorative molding often stops atop the baseboard molding.

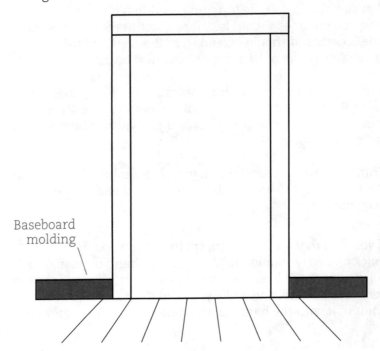

Baseboard molding

If you use narrow boards as outside corner molding, you can install them by using one narrow board and one slightly wider and nailing them in lapped fashion at the corners. If one board is 3 inches wide, nail it up so that the inside edge of the board is flush with the rough corner of the wall. When you have the first board nailed up, use a second board on the opposite side of the corner and let it lap the edge of the first board. To get a balanced look on the corners, the second board should be the width of the first plus the thickness of the first board.

When the two boards are installed in this overlapping manner, the balanced look is achieved. The rough or uneven ends of the corner boards are concealed, and the entire portion of the room is improved.

You can also buy corner molding for both inside and outside corners. This molding, which is lightweight, very workable, and somewhat fragile, can be installed with very small finishing nails to hold it securely.

The inside corner molding fits into the square corner and conforms to the wall lines readily. It is usually fitted above the baseboard molding and under the quarter-round, cove, or bed molding used at the juncture of ceilings and walls.

Outside corner molding is essentially the same. The only major difference is that it is manufactured with a reversed angle from that of the inside corner molding. The outside corner molding will fit over corner boards and can be installed exactly like the inside molding.

If you have cabinets, built-in bookcases, or other permanent wall irregularities, you can install quarter-round molding or inside corner molding around the fixtures to produce a more finished look. This molding is especially effective in dressing up the sides of cabinets and bookcases.

To trim exterior walls and windows, you will need to use the same basic approach as that used inside the house. The exterior window trim usually acts as a link between the window framing and the siding or exterior wall covering.

In many instances the trim is installed, as it is inside, and the boards or other wood or simulated wood covering then comes to the outside edge of the trim boards. The same is true of exterior doors.

You can install trim inside the window frame where bricks are used as wall covering, and this same type of trim work can be done at doors. This work conceals the often unfinished appearance of the windows and helps stop or slow the loss of heat or cooling.

Components of window trim are the stool, head casing, side casing, apron, and mullion. The *head casing* is the board or trim material installed over the window. Head casing joins the side

Installing corner molding

boards, called *side casing*. The trim material separating the units of double windows is called *mullion trim*.

Where the window sits on the sill you will install a part of the window called the *stool*. This wood unit rests partly on the sill and partly on a support piece under it, called the *apron*.

The stool helps to seal the space at the bottom of the window where the sash is in contact with the sill. You cannot install a window so tight that there is no entry space for moisture and cold or hot air. Such a tight fit would make it impossible for you to raise and lower the window sash.

The stool provides the weather barrier needed. The apron is a unit of wood similar in dimension to that of the stool. It is fastened under the stool and is nailed to the window framing.

In order of installation, the stool is installed before the casing is in place. The exterior casing rests upon the stool. The side casing is usually installed next, followed by the head casing. The apron is the last part of the window trim installed.

The apron is usually cut to the same length as the outside width of the side casing. That is, when installed under the stool, the apron will reach from a point directly under the outside edge of one side casing to a point directly under the outside edge of the other side casing.

The final step in installing window and door trim is to set or sink the nails, using a punch. You can fill the nail holes before you paint. If you do not fill the nail holes, you may find that unless you use special coated nails, the nail heads will rust, and rain will wash the rust stains down the wood surface, creating a displeasing appearance.

Soffit & fascia installation

THE FASCIA in a house is the material nailed over the ends of the rafters at the eaves. The *soffit* is the material installed under the rafters at the eaves. The purpose of these steps is to enclose the rafter ends and eaves from entry by rain, wind, and other undesirable elements.

These two steps are fairly simple and can be done in two basic ways. The first method uses bevel-cut fascia and horizontal-cut rafter ends. The other method uses fascia with straight-cut ends and rafters with the slant left unmodified.

Basic ways to install fascia

You can combine the two methods. You can use modified rafter tail cuts and bevel-cut fascia, or you can use uncut rafter ends and straight-cut fascia.

When you are finished with the work, you will have the rafter ends enclosed, and you will also have ventilators installed in the soffit. These ventilators allow air to circulate between the rafter insulation, if you used any, and the roof itself. If you do not permit the circulation of air so that overheated air can escape, you run the risk of damaging your roof and ceiling. This is in reference to cathedral-ceiling construction. The step is not nearly as important if you use insulation laid between ceiling joists rather than between rafters.

Start by deciding which method you will use. If you choose to use the bevel-cut fascia, select the boards and rip them or bevel-cut them across the top. Before you install the fascia boards, you will need to make the horizontal cuts on the rafter tails, unless you made the cuts when the rafters were installed. The purpose of the horizontal cut is to permit you to install soffit parallel to the ground or in a level fashion from the fascia to the ledger against the house.

Cutting rafter tails

You may also need to cut the rafter tails to make them narrow enough for the fascia boards to cover the ends. The best way to handle the job is to measure down from the bottom of the sheathing or top edge of the rafter to a point at least an inch shorter than the width of the fascia boards.

If you used 2-x-10 rafters, these will be 9½ inches wide. If the fascia board is 7½ inches wide, you will need to cut off at least 3 inches from the bottom edges of the rafters. Measure up 3 inches from the bottom edge of a rafter, at the very end, and mark the point. Use a level to get a reading for a horizontal cut line from the point you marked across the rafter tail. Mark the line clearly and do the same with all other rafter ends.

3"

Use a circular saw to make the cuts. This can be a dangerous operation, so it is better to make the cuts when the rafters are installed. If you did not, work with great care while using the circular saw.

When all the rafters are cut horizontally, you are ready to install the fascia boards. This job is much easier than the one you just completed. Use boards that are straight, sound, and acceptable in appearance. If the house front (or back) is 52 feet, you will need 54 feet of fascia boards (an extra foot on each end for roof overhang). The purpose of the bevel cut is to allow the board to fit snugly under the sheathing overhang. Bevel the top edge of the board so that the cut will slant downward from inside edge to outside edge.

Bevel-cut fascia

The board should be wide enough to cover the entire end of the rafter and have enough left over to hang an inch below the bottom edge of the soffit. This extra inch is the drip edge, which will keep water from running back under the soffit and damaging it.

Cut the fascia boards so that they will end on a rafter end. Do not butt-join them between rafters; it is difficult to keep the joint from separating and admitting moisture and insects.

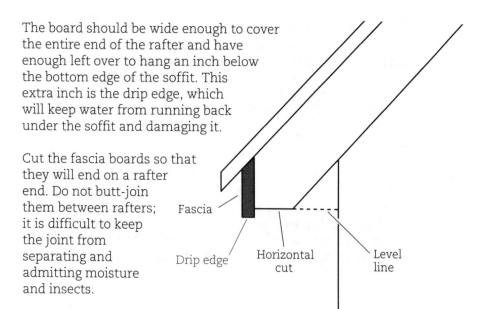

Fascia

Drip edge

Horizontal cut

Level line

If you plan to paint or water-seal the boards, it is a good idea to do it before you nail them up. Lay the boards on a good work surface and water-seal them on one side, let the side dry, then seal the other side.

You can lay out the fascia boards for front and back eaves. By the time one side is complete, the first boards will be dry enough for you to turn them and start the sealing process on the back side.

For the fastest method, use a garden sprayer and fill it one-third to two-thirds full of a water sealant. There are many good ones on the market. Pump the sprayer until you have enough pressure to cause a steady spray. You can spray an entire board in a matter of seconds.

If the weather is warm and sunny, the boards will dry enough in only a few minutes to allow you to install them. You will need help in holding the boards while you nail them. Have someone climb a ladder or stand on a scaffold and hold one end so that it is even with the rafter ends while you nail at the opposite end. Start by evening the board so that the end of it is flush with the outside edge of the final rafter. One or two nails

Painting fascia boards

will hold the end securely while you move along the eaves and drive at least two nails through the fascia board into the end of the rafter. Attach the fascia board securely to every rafter end.

When the board is installed, continue along the eaves until every rafter end is covered. Be sure to turn the bevel cut so that the board can fit all the way up into the space under the sheathing.

If you choose to use straight-edge boards, you will do the work exactly as described above, but the fit under the sheathing will not be as tight. You will still want to allow the inch of extra width for the drip edge on the fascia board. An astonishing amount of water can cling to the bottom of the board, work its way into the soffit, and do considerable damage in a brief time.

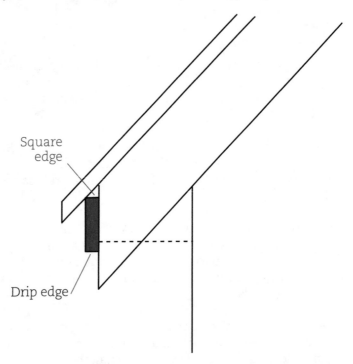

Square edge

Drip edge

Preparing for soffit installation

When the fascia is installed, you are ready to start on the soffit, the wide expanse under the rafter ends extending from the ends of the rafters to the siding of the house and continuing along the entire length of the wall. These are two ways to install this material, which often consists of plywood or special sheathing materials rather than ultrawide boards.

If you horizontal-cut the rafter ends, you will need to install a ledger against the wall along its full length. This ledger is simply a plate to which you can nail the soffit.

Use a level, either a regular carpenter's level or a line level, and establish and mark a level line from the bottom edge of the horizontal cut of the rafters. Do this at each end and at the center of the wall. Mark the level points as you establish them.

Using a chalk line, strike a guide mark along the length of the wall to connect the three marks. You should have someone hold one end of the chalk line on one mark at one end of the wall while you pull the line to the other end of the wall and hold it on the final mark. When you snap the line, it should cross the middle mark and make a perfectly level line along the whole wall.

If the chalk line misses the mark at the center, establish your level lines again to double-check. If you are certain that the level lines are correct and the chalk line still misses the middle mark, you know that there is a problem with the rafter line at the eaves. It is not likely that there is a major discrepancy; however, you can tell from a visual inspection that the alignment is incorrect if the error is great.

Soffit installation

When the line is ready, nail up the ledger plate, which can be 2-x-2 or 2-x-4 stock. The bottom edge of the ledger plate should be aligned and flush with the chalk line.

You can now cut soffit material and begin to nail it in place unless you need to install vents in the soffit. Remember that if you use a roofline for a cathedral ceiling, you will need to use vents.

Vent installation

You can buy light metal vents that are very easy to install. Lay the vent upside down on the back side of the soffit material and mark around the vent itself, not the flange that extends to

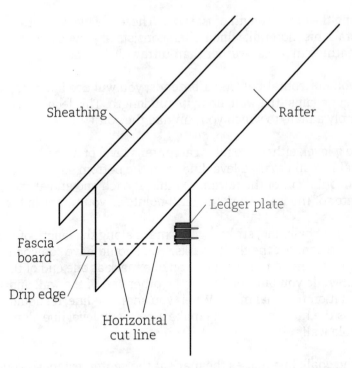

Sheathing

Rafter

Ledger plate

Fascia board

Drip edge

Horizontal cut line

the side all around the vent. Drill a hole along the line so that the outer edge of the bit barely touches the line. Then use a jigsaw or keyhole saw to cut out the elongated oval you marked around the vent.

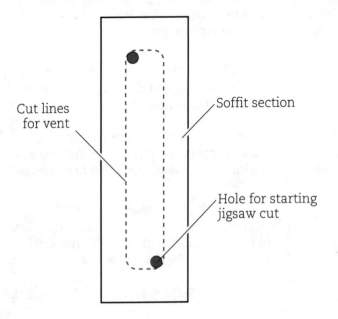

Cut lines for vent

Soffit section

Hole for starting jigsaw cut

If you can handle a circular saw effectively, you can use this faster cutting method for the long sides of the cutout mark. Use the jigsaw or keyhole saw for the oval cut lines.

With the back side of the soffit material still facing upward, set the vent in position so that the flanges on the sides lap over the soffit surface to keep the vent from falling through. Now, assuming that the soffit material has already been cut, lift the section into place and nail it securely to the horizontal cut of the rafter ends and the bottom of the ledger plate.

Do this along the complete length of the house. Be sure that all section ends rest against a rafter as well as the ledger plate; otherwise, the material will sag after a short time.

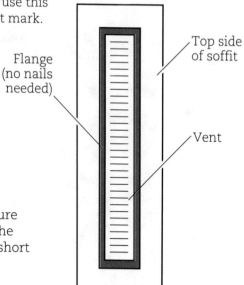

Flange (no nails needed)

Top side of soffit

Vent

Completing soffit installation

If you do not need to install vents, you can cut soffit sections and nail them in place as a solid unit. Measure from the side of the wall to the inside edge of the drip edge and mark and cut the soffit material accordingly.

It was mentioned earlier that there is a second way to install the soffit, a way that does not involve the horizontal cut on the bottom of the rafter tails. You simply let the soffit follow the slant of the rafters to the point where the rafters cross the top plate or top cap of the wall frame.

Measure from the wall of the house at a point just under the rafters to the fascia board. Cut the soffit material to fit and nail it in place.

The best reason for using this second method relates to the pitch or slant of the roof and the amount of eaves or overhang. If your roofline is very steep and you have a fairly wide overhang, the roofline drops so sharply that a horizontal line from the bottom of the rafters to the side of the house may actually be lower than the tops of the windows. If such is the case, you will need to install the soffit in a slanted fashion. You

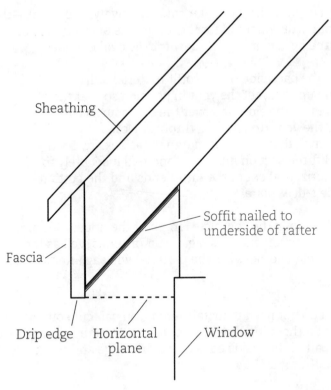

Sheathing

Soffit nailed to
underside of rafter

Fascia

Drip edge   Horizontal
plane

Window

can nail to the bottom of the rafters all the way, and you will
not need to use a ledger plate.

When the fascia and soffit are complete, you are ready to paint,
stain or water-seal the surface of the materials and then install
the guttering system along the front and back of the house. The
simple way to do this is to measure to the midpoint of the front
or back and mark the point near the very top of the fascia
board. Using a line level to guide you, run a line from the mark
to the end of the house. You will need to mark the exact level
line first and then determine the amount of drop or fall you
need for the guttering.

Strike a chalk line from the center point to the point where you
decided the fall line should end. You can measure 1 to 2 inches
lower than the level line, then chalk the line to the lower mark.
This drop will assure you that water from the roof will drain
down the guttering without difficulty.

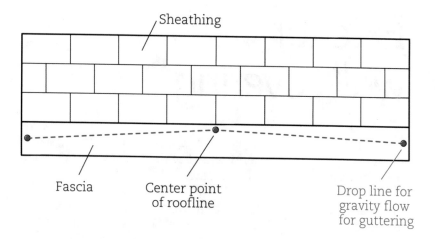

Sheathing

Fascia

Center point
of roofline

Drop line for
gravity flow
for guttering

When you install the guttering, make certain that the top of the
guttering conforms to the chalk line you made as a guide. You
can then add the downspouts and the remainder of the
guttering.

Exterior wall covering

EXTERIOR WALL covering is one of the most important parts of the house. The entire exterior effect of the house depends upon the appearance of the wall covering.

Clapboard or board siding is highly attractive and may be installed easily at a reasonably low cost. The work is light, and installation is rapid. One of the best characteristics of board siding is that you can stop whenever you feel like it. If you are using masonry materials, you feel an obligation to use all the mixed mortar, and you must scratch or joint the finished work, brush it, then spend some time cleaning up the mortar box and your tools. With wood siding, you can stop with a nail half driven if you wish.

When you order boards for siding, you can compare prices of 1-x-4, 1-x-5, 1-x-6, and other boards. Ask the dealer to give you a rough estimate of how many board feet will be needed to cover the house, then get comparative prices of the various dimensions.

When the choice is made, set up a work area that includes sawhorses or sawbucks, extension cords, the basic tools needed, plenty of nails, a square, a tape rule, and your lumber. Before the job is finished, you will need scaffolding or ladders.

Installing corner boards

You need to install corner boards first. These are straight, defect-free, and attractive boards that will frame the corners of the house. Use one 1-x-4 and one 1-x-5 board on each corner. The board should reach all the way from the top of the foundation wall to the top of the wall.

Nail up the 1-x-4 first. Let the outside edge of the board align perfectly with the corner of the house. Use coated nails to fasten the board to the corner post of the wall framing. Nails should be spaced 1 foot to 18 inches apart.

The second board is the 1-x-5 stock. Let it lap the edge of the first board so that the outside edge of the 1-x-5 is aligned with the outside edge of the 1-x-4. When the second board is installed, you will have a waterproof and insect-proof corner that will not only prevent a great deal of heat and cooling loss but also provide a neat and unobtrusive point for your siding boards to end.

1 x 4

1 x 5

Install these corner boards at every exterior corner of the house. When all corners are covered, you are ready to install the siding.

Nailing up drip caps

Start by installing a drip cap. This is a 1-x-1 or 1-x-2 strip nailed against the header at the point where the foundation wall meets the wall frame. The purpose of this drip cap is to divert water from the foundation wall and let it drip onto the ground.

The drip cap will extend from the inside edge of one corner board to the inside edge of the next corner board. Nail it in place so that it covers the crack between wall frame and foundation wall.

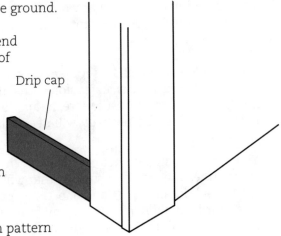

Drip cap

Patterns for installation

Plan your installation pattern before you begin to nail up siding boards. When you order your lumber, you should keep in mind that you can reduce waste to a bare minimum if you plan correctly and cut your sawing and measuring time.

Remember that the distance from one stud to the next is 16 inches on center. A 12-foot board will cover 10 studs, including the corner post. An 8-foot board will cover 7 studs, including the corner post.

If you establish a pattern of using a 12-foot board followed by other 12-foot boards in a house 52 feet long, you will need four full-length boards and a 4-foot board. Each of these boards will end on a stud, which is the way you want it to work out. Do not let a board end anywhere but on a stud. On the next course, let the first board run 8 feet, which will cover the first seven studs. The next boards will be 12-footers; each will end on a stud and lap the end of the previous board by 4 feet.

By following this pattern, you will be able to stagger the ends of the boards effectively, and you will have virtually no cutting or waste. Never let two boards of adjacent courses end on the same stud.

Your pattern will be one course of 12-foot boards, to be ended by a 4-foot board, alternated with a course starting with an 8-foot board, changing to 12-foot boards, and ending with an 8-foot board. When you cut an 8-foot board to end the first course, you have 4 feet left. You will use this 4-foot board on the next 12-foot course.

Whatever the dimensions of the house, you can usually work out a pattern that will be effective and efficient. The less cutting and measuring and the least waste will mean huge savings to you in time, energy, and money.

When you install your first board, drive three or four nails just under the drip cap. Do not sink the nails all the way. Lay the first board, a 12-footer, so that it rests upon the nails. Use your level to be sure that the board is resting in a true horizontal position. Do not start a wall with a board that is not level and true.

Starting to install siding

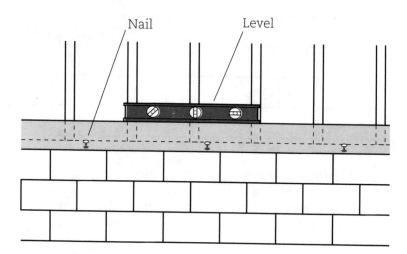

Nail   Level

If the reading is satisfactory, butt the end of the board to the side of the corner board and make sure that the bottom edge of the board covers the entire drip cap. Nail the board by driving nails through the board into every stud along the path. Do not try to nail too low on the board because of the drip cap.

If you wish, you can have an entire wall with few if any nails showing. Keeping nails away from the weather helps prevent rusting and staining of the boards.

Continue the course by butt-joining the second board against the first. Nail as before, and keep the progression moving until you have covered the first course of the wall all the way across.

When you have completed the course, you can start the next course by marking the lap line. You will want to let the top board lap over the top of the bottom one by at least 1 inch.

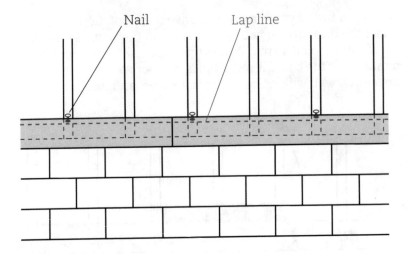

Measure down 1 inch at the beginning, middle, and end of the board length and set the board so that it rests on the nails and laps the lower board correctly. Butt the end of the board into the corner board and prepare to nail.

Making and using a story pole

You can pause to construct a story pole, or you can buy one. A homemade one works well and cost very little. You can use the back side of an 8-foot board and later use the board in the wall, and the story pole will cost you nothing.

Stand the pole so that one end rests on the footing surface beside the foundation wall. Mark where the story pole strikes the bottom of the first board. If you are using 5-inch boards, lay off the remainder of the story pole in 5-inch gradations. Because a 5-inch board is actually 4½ inches wide, lay off the story pole in 4½ inch increments.

Each time you start to nail up a board, use the story pole to be certain that the end of the board conforms exactly with the mark on the pole. Check the other end to be sure there is no drop along the way. Each time you nail up a new board, make the same checks at both ends.

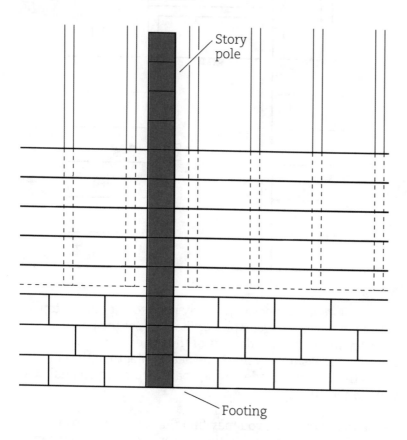

Story pole

Footing

Occasionally a board will not be exactly the same width as other boards, and your reading will be slightly incorrect. You can make minor adjustments by lapping slightly more or less with the next board. When you encounter wall interruptions such as doors and windows, use a line level as well as the story pole to be sure that you start the board on the opposite side of the window or door at the proper height.

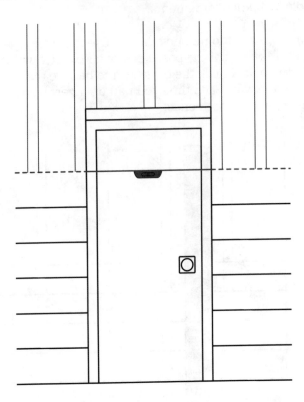

Installation around windows and tops of walls

When you reach windows, you may need to cut out the top or bottom of a board to get a proper fit. If so, set the board in position, mark the outside edge of the window frame on both sides, and measure and mark the needed depth of the cut. When the cutout is made, fit the board into place and nail it as you did the others.

At the top of the wall, you may find that you must rip an inch or so off a board to fit it into the remaining space. If you can tell in advance that such a problem will occur, you can make minor adjustments on the lapping of boards and end the wall with a perfect fit.

At the top of the wall, finish the work by nailing up *cove molding*. This molding conceals the juncture of the wall covering and the roofline and prevents moisture and insects from getting behind the siding.

One of the delightful aspects of wood siding is that you can buy many styles of siding, and when you are ready to work, you can arrange the siding in a variety of patterns. You can install it horizontally, vertically, or diagonally, or you can use a combination of patterns.

When you install board siding in a vertical fashion, you will need to use *batten*, or thin strips of wood, to cover the cracks and spaces between the boards. Use wide boards alternated with more narrow ones for interesting effects. If you have several widths of wood, try not to use any two of the same width beside each other.

Putting up vertical siding

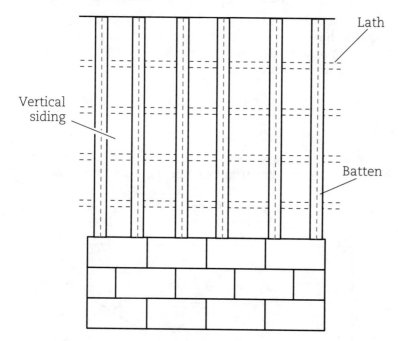

Lath

Vertical siding

Batten

You may need to install lath behind the boards as well as over the cracks. Remember that you will not have a nailing surface behind the wood except when you are over a stud. You can use 1-x-2, 1-x-3, or 1-x-4 lath or nailers. Run these across the top, bottom, and center of the wall at least. You can also run them one-third of the way and two-thirds of the way to the top of the wall in addition to top and bottom areas.

On narrow boards (4 to 6 inches) use at least two nails wherever the board crosses a nailer. For wider boards, use at least three nails at each nailer crossing.

You can avoid waste in vertical board siding if you buy boards the exact height of the walls. If these are not available, you can buy different lengths so that in combination the boards will work out to waste-free coverage or nearly so.

If the wall is 9 feet high, you will have at least a foot of waste, because boards are typically not available in 9-foot lengths. You may have to use 10-foot lengths and cut off the excess foot. Later you might find a use for the scraps. If not, they can always be used as firewood.

Do not burn treated lumber scraps. These give off toxic fumes that can be very dangerous.

Installing diagonal siding

When installing diagonal siding, divide the wall into equal parts by striking a chalk line from one upper corner to the opposite lower corner. If you need to, keep measuring and marking off wall space, 3 feet at a time, from the diagonal line until you know exactly what incline to use for the boards.

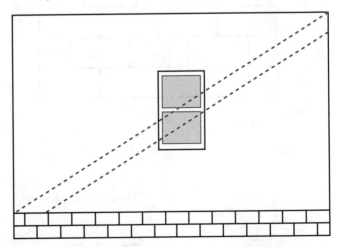

When you start nailing up boards, nail where the boards cross studding and eliminate the need for nailers.

Tongue-and-groove 2-x-6 lumber makes exceptionally good siding. Start with the tongue facing down at the top corner of a wall and follow the diagonal slant downward. With each course, position the groove into the tongue directly above it. In this fashion you can use the tongue space for nailing so that no nails will show in the wall.

When you reach the bottom of the wall, you will still have room to nail. The height of the foundation wall will provide nailing space. If you start at a bottom corner and work your way up, you will not be able to nail at the top of the wall.

It is usually better to choose a less conspicuous wall for your early efforts. A back or side wall will be a good place to learn how to do your best work.

You can buy several types of simulated wood or hardboard panels that are installed as you installed the board siding. Some of these panels are only 5 or 6 inches wide; others are 1 foot wide. Nail them in place, lapping as before, and use a story pole to keep courses aligned.

Putting up simulated wood

You can install full-size panels of plywood as exterior siding. Some panels are smooth-surfaced; others are scored and marked to simulate boards of varying widths. To install these panels, drive the nails partway into the crack between the foundation wall and wall frame, as you did before, and position the panel as you want it installed. Use a level on the side of the panel to be sure of a true vertical alignment.

You can align the side of the panel with the corner post if you wish. A 4-foot panel will end on studs for convenient nailing. You can use corner molding to hide the joints when the wall is complete.

The edges of plywood panels are often manufactured for effective joining with either a lap or a tongue-and-groove effect. When you start a wall, place the inside lap on the edge of the panel away from the corner. When the next panel is installed, the outside lap will fit over the inside lap and provide an effective waterproofing joint. If the panels have straight edges, you can buy lath to install over the joints.

When walls are high enough so that you will need more than one panel in height, use similar lath to install horizontally over the joints. You will need to cut out for windows and doors as you install the panels, and this is usually a very simple matter if the window frames were installed correctly.

One easy way is to set the panel in position except that you butt the edge of the panel against the window frame. Mark at the top and bottom the point where the cutout should be made. Then take down the panel and measure from the edge of the last panel installed to a point 48 inches away. Mark the point, then measure from the point to the outside edge of the window frame.

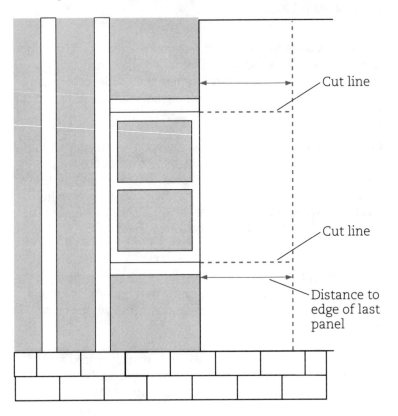

Cut line

Cut line

Distance to edge of last panel

Assume that the distance is 17 inches. Start at the top mark on the panel you used earlier and measure in from the outside edge of the panel 17 inches. Use a square to mark a straight line, at right angles to the edge of the panel, to the end of the cut line 17 inches into the panel. Do the same at the bottom mark. When all marks are made, use a chalk line to strike a line from the end of the top line to the end of the bottom line.

Cut out along the lines. You can use a circular saw for the middle portions of the inside cut, then use a handsaw or keyhole saw for the final part of the cut. Use a circular saw for the two cuts from the outside of the panel. Use a handsaw when you get close to the point where the two cuts meet.

When you put up shakes or shingles, follow the same principles that you used in roofing. Remember that the first course is doubled for complete watershed protection, and if you are using synthetic shingles, you may want to install the first course upside down so there will be total watershed action.

You must determine how many inches there will be "to the weather," the amount of the shingle that will usually be exposed to sun, wind, rain, and other weather. You must lap half a shingle in each course.

With wood shakes and shingles, you can leave a small space between shingles to permit swelling from heat. Shingles need at least ¼ inch to expand. You can leave a full inch if you wish.

It is a good idea to use a story pole idea on any wall-covering work that involves courses. The story pole can keep you from the tendency to let one end of a course "float," or start to rise higher and higher until the end of the wall will need extensive reworking before it is acceptable.

Vinyl siding has become extremely popular in recent years. This weather-resistant material does not decay, rust, or weaken when exposed to weather. It will last for decades and look as good as it did when installed.

Putting up vinyl siding

One of the finest properties of the vinyl is that you never need to paint it. It is essentially maintenance-free, and this advantage substantially reduces your long-range expenses.

Vinyl is also easy to install. Follow the directions printed on the products. There are several methods of installation, and you must follow specific installation procedures for the warranty protection to be in effect.

For brick and cement-block wall coverings, refer to *Rough Carpentry Illustrated, second edition* (#4395). TAB also has other titles devoted to masonry work.

Bookcase wall covering

CERTAIN ROOMS in the house need more appointments than the typical den or bedroom, which may have walls covered with panels of decorative plywood or paneling, wallpaper, wallboard and paint, or boards. A personal library or study might well require walls covered with bookcases.

Many do-it-yourselfers stay away from complete wall bookcases because of the expense and time needed to construct the shelves. The truth is that such full-wall bookcases are not difficult or expensive to build.

Your first concern is what to do about wall receptacles and light switches. These are not major problems. You can simply enclose them in the bookcase area and even construct a small compartment for the switch and receptacles.

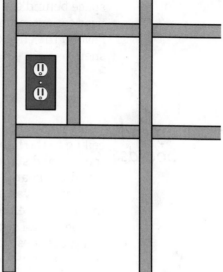

Planning construction

When you are planning the bookcases, plan to cover the entire wall from floor to ceiling. It is easier to cover the wall completely than it is to interrupt the flow of the work.

You will find that if you build a bookcase on the floor in the room, you will not be able to stand the case into its final position. You are better off to build it in the place it will occupy.

Start by installing one end of the case in a corner so that the board will rest against the adjacent wall. Many people use 1-x-12 or at least 1-x-10 lumber. This can be a very costly mistake. Most books are not that large. You will find that a

shelf 8 or even 7 inches wide will
hold nearly all the books on the
market today, and 8-inch shelving
can be bought for far less than 12-
inch boards.

You can use 6-inch boards and
make a bookcase large enough to
hold the *Encyclopedia Britannica* and
similarly oversized books. You
simply install the first boards an
inch or two from the wall. The
space behind the shelves can be
used for extending the books
beyond the edge of the
shelf.

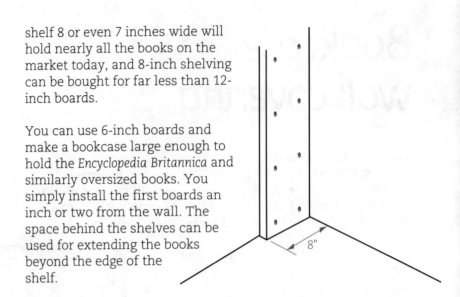

Installing end boards

When you put up your first end board, measure from the
ceiling to the floor and cut the board 1 full inch shorter than
the actual distance or length.
Stand the board in place
against the wall adjacent to
the one where the bookcase
will be built. Let the board
stand 2 inches from the
corner. Fasten the board to
the wall by using nails or
screws into the studding in
the corner where the
partition wall was formed
when the house was
framed. Do the same in
the opposite corner.
Leave the board
1 inch too short
again.

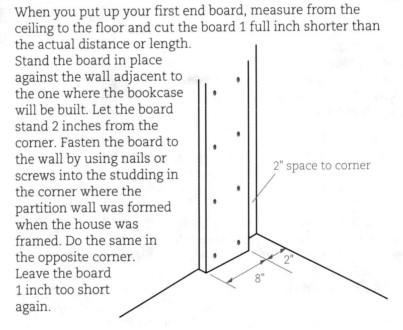

When both boards are installed, measure from wall to wall and cut a board that length. Slip the board over the top of the boards already installed in the corners.

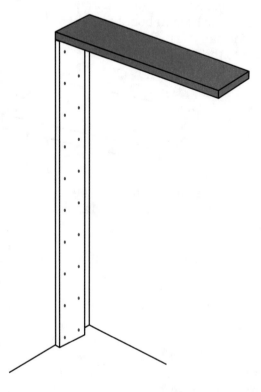

You cannot nail the top board in place, so you will need to use corner braces. Use two 1-inch braces to fasten the top board to the side board. You can also use screws or nails to fasten the top board to the ceiling joists so that when filled with books or other display items, the bookcase cannot topple or be pulled over.

Keep the bottom shelf off the floor at least 3 inches. The thickness of a 2-x-4 stood on edge is almost perfect, so you can cut enough blocks of 8-inch 2-x-4s to reach across the whole floor along the wall when spaced 2 feet apart.

Installing top and bottom boards

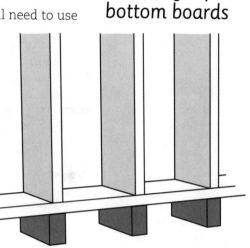

Cut another board long enough to fit
between the two end boards and lay it
across the 2-x-4 blocks. Use 1-inch corner
braces to attach the bottom board to the
end boards. You can use nails to attach the
bottom board to the 2-x-4 blocks.

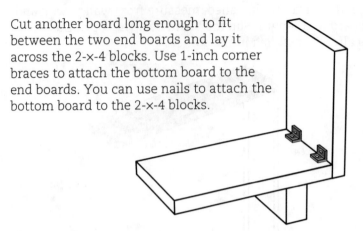

Installing upright boards

Next install upright boards 24 to 36 inches apart. Run these
boards from the top to the bottom board and fasten them with
corner braces.

Decide now how high you want the individual
shelves. You can buy shelf fasteners or shelf
holders and leave the spacing decision until you
use the shelves. These shelf holders are strips of
aluminum with slits for shelf brackets. You
fasten the aluminum strips up and down the
upright boards. Use two strips on each board,
locating the strips about 2 inches from the edge
of each board. Use screws or nails to hold the
aluminum strips in place.

Fasten the brackets onto the
bottom of the shelves, then
insert the free end of the
brackets into the spaces in the
metal strips you have already
installed. By using this manner of installing shelves, you can
move shelves as you wish without trouble or tools.

A good idea is to set up the shelves at varying heights so that
you can use the shelves for videotapes, books of varying sizes,
coffee-table books, and sets of encyclopedias. The 1-inch-thick

shelving boards will be strong enough to hold a 2-foot to 3-foot shelf of books.

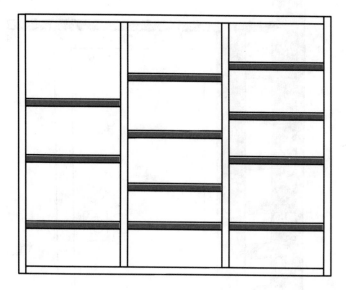

If you make the shelves all the same width, you can move the shelves interchangeably across the entire room. This will give you much greater flexibility.

Be sure to make use of the space above doorways in the room. Let the top board, installed against the ceiling, reach over the doorway and into the corner. The top shelf can run across the door casing, and you will be able to make full use of every inch of space in the wall.

Even the space behind the door can be used. Run boards up and down the wall and against the door facing. You can install shelves 6 inches long and use them to store paperback books or less expensive items.

Utilizing space better

Where there are switches or receptacles, install the full-length upright boards so that they run close to the switch or receptacle box. Then install short upright boards 6 inches from the full-length boards on the other side of the switch or

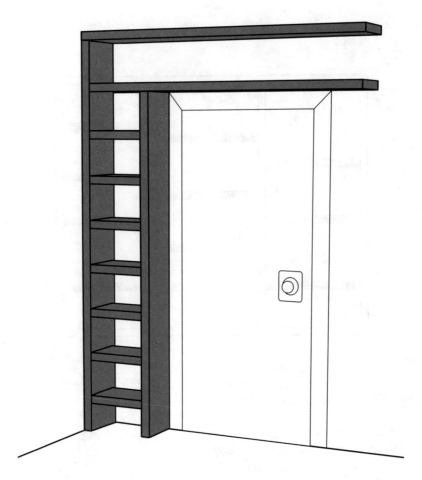

receptacle. When you use the plugs, you can run the cord into the small cubicle.

Using wood strips for shelf installation

If it is not convenient for you to buy the aluminum strips, you can install the shelves quickly and easily by using short shelf-holder strips that you can cut from scrapwood. Earlier it was mentioned that you might have foot-long boards left over from another job. You can now make use of the scrap lumber.

Use a C-clamp to hold the foot-long lumber steady so you can saw it. Use a circular saw to saw off strips 1½ inches wide. Shorten these strips to 6 or 7 inches.

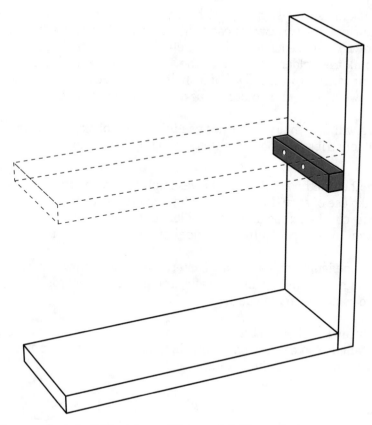

Use an electric drill and small bit and drill two holes in each strip. One hole should be a third of the way from the end and the other should be two-thirds of the way from the end. Measure up from the top edge of the bottom board to the point representing the height of the shelves you want. Assume that you want the shelf to be 10 inches high. Mark the point on both upright boards and use your square to mark a level line from the front to the back of each upright board.

Start screws 1½ inches long through the holes you drilled. When the screws are through the narrow shelf holders, position the strip along the marked line and fasten it to the upright board.

Do this on both sides. Then cut a board to the proper length, insert it between the upright boards, and let it lie across the shelf holders.

You can use a very small finishing nail to fasten the shelf to the shelf-holder strip, or you can drill a small hole and use small screws, one in each end of the shelf, to hold the shelf in place. Install the holder strips for all the shelves you need. When you want to change the height of a shelf, you can remove the screws holding the holder strips and reposition the shelves.

You will find that you cannot use full-height books under the holder strips. of wood. Use this space for shorter books.

Using upright shelf supports

Another method of holding shelves in place is yet another way to make use of scrap lumber. If you have short boards 6 or 7 inches wide, cut the scraps to the proper length for shelf height, then fasten the scrap lengths to the sides of the upright boards.

Use one length on each side. Fasten these with screws short enough not to reach all the way through the two boards. Two screws per board will suffice.

Now lay the shelf on top of the ends of the boards you have just installed. You can use small finishing nails, screws, or corner braces to hold the shelves in place.

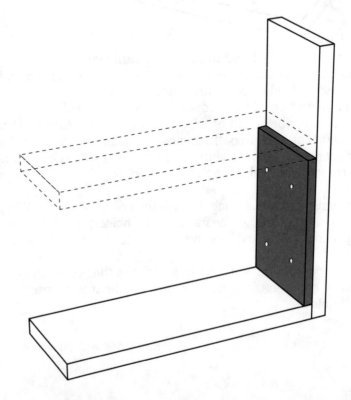

If you do not want to cover a complete wall with shelves, you might cover the space between windows or between a window and a door. If there is a small alcove that has no particular use, you can install shelves and convert the alcove into a bookcase or novelty case.

You can use the same basic plan to install shelves to hold a VCR, stereo equipment, tape decks, or television sets. The space not occupied by the equipment can be used to hold records, tapes, and similar items.

If you have a desk along one wall, you can install shelves above the desk so that the space on the upper part of the wall will not be lost or wasted. Because of the suspended weight of shelves filled with books, you will want to incorporate extra holding power.

Use a full-width shelf board across the back of the shelf assembly. The board will fasten to the back edge of all the upright boards and will in turn be attached to the wall studs, so that the bookcase will be supported by the board, ceiling joists, and studding.

You can build a small and very simple desk by using upright sections of plywood with a longer panel across the tops of the upright sections. Use a section of plywood 30 inches wide; later you will fasten this to the wall in a corner. Cut another segment the same size and set it aside for the moment, along with the first piece. Now cut a length of plywood 40 inches long and 30 inches wide and install it to cover the section already fastened to the wall and to the second piece you cut and set aside. Use a fourth piece to form the back of the desk. This piece will be 40 inches long and 30 inches high.

Use wood glue and nails or screws to unite and hold this assembly. Now set the assembly into a corner and fasten one of the panels to the wall in the corner. The remainder of the desk will extend along the wall.

Other shelf construction

You can now build bookcases or shelves from the free side of the desk to the wall. The plywood top of the desk can be double thick if there will be a great deal of weight on it.

You now have a very simple and inexpensive desk area that works well for holding a computer or typewriter. You can run a full 8-foot length of plywood along the wall and use another section of plywood to act as a support for the extended desk surface. This surface can hold a printer, pen sets, desk calendars, and similar items. You can use the basic ideas presented here to build a number of other wall-covering items that are functional and inexpensive.

Installing flooring

ONLY WHEN the rough and dirty carpentry work is done should you give any thought to installing finish flooring. When you framed the house and completed the rough carpentry, you had subflooring for a walking and working area.

On the stairways, you had rough treads because the finish treads would be scarred and dirtied by muddy feet and rough materials dragged over them. These can also be replaced by finish treads.

This chapter deals with the materials associated with do-it-yourself permanent installation of floor coverings. You have a choice of materials, including tile, vinyl, wood, linoleum, ceramic, brick, slate, flagstone, and a lengthy list of other flooring materials.

Wood, including boards and tongue-and-groove flooring, has been among the most popular flooring materials for more than a century. These flooring materials are easy to install and cost relatively little. They can be coated with polyurethane, stains, wax, or paint; they last a long time and are not highly susceptible to damage from ordinary usage.

Installing wood flooring

Boards are not as popular as they once were, primarily because tongue-and-groove flooring is so easy to install and the final appearance is better than that of the usual board flooring. If you use boards, you should be careful in your selection and accept only those boards free of noticeable defects and warp or twist.

Use boards no thinner than 1 inch. Width is no problem as long as you stay with 5-inch and wider widths. If you can find 12-inch boards, these work extremely well and reduce the time for installation.

Installing flooring   185

Making a square start

Installation is extremely simple. All you need to do is get a square start in one corner and against one wall, then proceed across the room.

Before you lay one board, install black paper or building paper over the subflooring. Completely cover the floor with the paper. This extra step, which takes very little time and money, helps prevent the floor from squeaking later when the floorboards settle against the subflooring materials. Before you install the black paper, you might need to walk back and forth across the subflooring where you plan to work. If you feel the subflooring give too much or you hear it squeak, add more nails.

Make a visual check of the subflooring too. Use shanked nails every 6 inches if you want a completely silent floor. The nails should be placed along the edges and over floor joists as well as in the center sections of the subflooring.

When the nailing is done, install the paper. Be sure to lap the paper at least 4 inches when you start a new course. Then you are ready to begin installing flooring.

The square start is important. Use a square in the corner to see that the corner is true. If there is a problem, modify the edge of the board that will be placed against the wall. You may find that you can correct the problem simply by modifying the end cut of the board. You may need to rip a thin strip from one edge.

Measure out from the wall the width of one board and mark the measurement at the corner and at the point where the board would enter, preferably at the wall at the other end of the board. Strike a chalk line to connect the two marks. You will not be able to align the outside edge of the board with the chalk line if the wall edge of the board does not fit properly. Make the necessary corrections.

Making boards conform to corners

Another simple method of getting a good start is to lay the board in position, and if it doesn't conform to the wall line as it should, make certain the end is cut perfectly square, then lay the board back in place with the square end against the wall.

If one end of the board does not touch the wall parallel to it, measure the distance from the wall to the edge of the board at the end away from the corner. If this distance is 1½ inches, for example, measure 1½ inches from the end of the board in the corner.

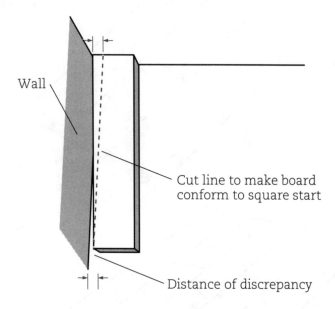

Wall

Cut line to make board conform to square start

Distance of discrepancy

Mark the spot, then stretch a chalk line from the mark to the other end of the board. Let the line stop at the point formed by the end and edge of the board. Cut along the line, and the board should fit.

It is better to use full-length boards in smaller rooms so that no end-joining will be necessary. At all butt-joined ends, if the two cuts are not exactly right, there will be a small but noticeable flaw in the flooring. This small gap is a perfect spot for dirt and small debris to collect.

If you must use shorter lengths, do not let any two boards in adjacent courses end at the same location. If necessary, use a long board in the center of the room and two shorter lengths at each end of the long board.

Use *shanked nails* to hold the boards in place. On wide boards, use at least three nails every 2 feet. For more narrow boards,

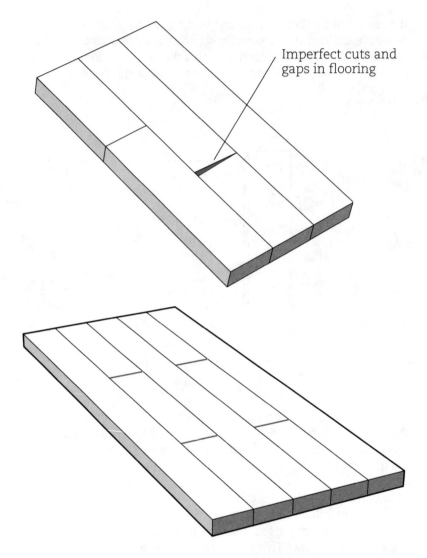

Imperfect cuts and gaps in flooring

use only two nails every 2 feet. Even if the nails have heads, you can use a punch to set them slightly and cover the heads with wood putty if you don't want nail heads visible.

Installing tongue-and-groove flooring

One of the most popular forms of flooring is tongue-and-groove *strip lumber*. It is made of very thin strips of wood, usually 2 inches or so wide, installed from one wall to the other wall of the room.

Get started with a square corner and a good fit against the wall. The wood strips are end-matched, so there will always be a perfect fit when two lengths are butt-joined.

Most modern wood flooring is undercut so that there is a slight oval cut into the bottom of the wood. This oval permits the strips to fit over slight irregularities in the subflooring without bucking or humping.

You can buy wood-strip flooring in hardwood varieties such as oak or maple. If it is tongue-and-groove lumber, like most modern flooring, you need not worry about the heads of nails showing.

One extremely satisfactory form of wood flooring is the tongue-and-groove board, available in 5-inch and 6-inch widths. You can buy the flooring in lengths of 12, 10, and 8 feet. You can sometimes find even longer lengths at supply houses or lumber yards with large inventories.

To lay the wide boards, start at one wall and lay the first board so that the tongue is facing into the room. Fit the first piece into the corner and against the wall as snugly as you can. Angle-nail the board to the floor by driving a cut nail into the tongue and slanting the nail head away from the board so that the point of the nail is headed toward the near wall.

Nailing boards properly

Note that the "point" of the nail is actually nearly flat. It is blunt rather than sharp. Remember that a pointed nail will split wood far more readily than a nail with a blunt end.

Use a cut nail every foot or 18 inches along the length of the board. Do not nail closer than 3 inches to the end of the board at either end.

Do not try to drive the nails in too deep. Stop before your hammer damages the wood. Later you can use a punch, laid flat, to set the nails completely. Hold the punch by the tip and lay the scored section across the nail head and parallel to the wood strips. Strike the punch until the nail head is flush with the surface of the tongue.

When you fit the next board against the first one, hold the outer edge of the board 2 inches or so above the subflooring and wedge the groove over the tongue. You may need to lay the board flat and tap the joint together.

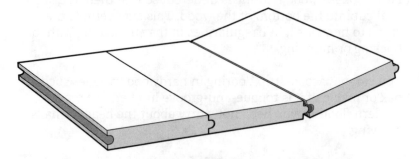

Do not hit the other board directly. The best method is to use a 2-foot length of scrap flooring placed so that the groove fits over the tongue well. Then use an ordinary hammer to hit the back side of the scrap board and drive the new board into position.

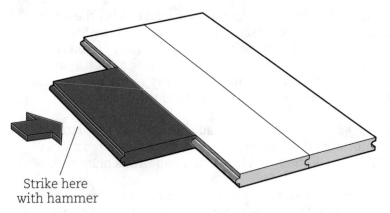

Strike here
with hammer

Correcting warps or irregularities

If this does not work, there is another method that rarely fails unless the board is hopelessly warped. Nail a short length of 2-x-4 to the subflooring about 6 inches from the board being installed. Then place the short length of scrap flooring in position, tongue and groove fitted together.

Now use a pry bar or other leverage tool (you can use a length of 2-x-4 or similar lumber) and with the lever between the scrap board and the 2-x-4, pry the board into position. If the tongue and groove are still not properly aligned, you may need to lay a scrap length of wood over the joint of the two boards and tap downward on the wood while prying.

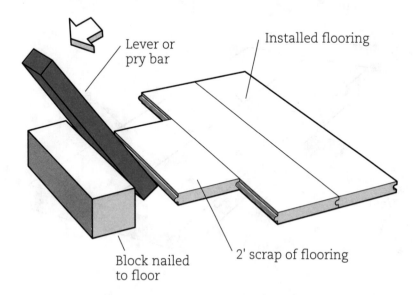

Lever or
pry bar

Installed flooring

Block nailed
to floor

2' scrap of flooring

As usual, if you need to join sections of flooring, do not let two boards in adjacent courses end at the same place. It is better if you can work out some pattern so that you can keep waste at a minimum. You will hear that you should allow 10 percent for waste, but this is true only if you do not plan and work accordingly. When we built a 520-square-feet deck, we had one length of 2-x-2 lumber a little more than a foot long left over, and this was not wasted. We found a use for it on another project.

You must cut out for all heating and cooling vents, and you must fit the flooring against cabinets and irregularities in the floor surface. When you reach vents, lay the board in position and mark the location of the vent opening. Remove the board and measure from the last installed board. Use a square or straightedge to mark the cutout.

Flooring around vents

You will have a small amount of free play around the vent. The metal flange will permit a ¼ inch or so of bad fit without the discrepancies showing. Fit all the floor lumber needed around the vent hole, and when the work there is done, insert the vent. Do not leave it open because there is always the danger that small tools or other items can fall into the hole, and the vent can be damaged.

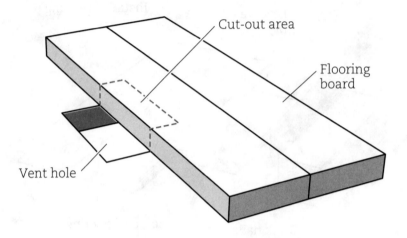

When you reach the other side of the room, you may not be able to nail the final boards through the tongue. Use a small finishing nail instead or let the baseboard molding hold the final edge in place.

Nailing down final boards

For the final boards, you may also find it necessary to saw or trim off the tongue for the board to fit into the space left. You may also need to trim the entire board slightly to get a good fit.

If the final piece nearly fits, you can sometimes place the board in position and let the final or outside edge lean against the wall slightly. Using a scrap of wood, set it against the flooring and tap downward slightly until the board is in place.

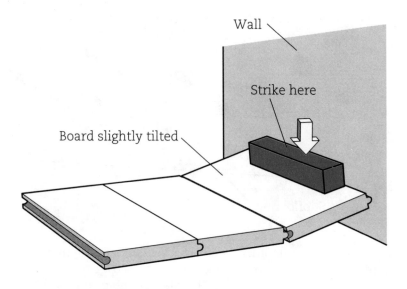

Wall

Strike here

Board slightly tilted

Pegged flooring

If you use wide boards, it is popular to bore shallow holes and use short dowel pieces to plug the holes to give the impression of pegged flooring.

You can have pegged flooring if you want to invest a little time to get the job done. Pegs hold extremely well and last a long time. Many houses that were pegged together before the Civil War are still holding together better than nailed houses.

To peg floors, drill holes in the boards, then into the subflooring. Try to hit floor joists whenever possible. Use a dowel the same size as the drill bit and cut a short section from the dowel. Coat the short section with glue and tap it into the hole. Clean away excess glue immediately.

Such pegging is highly attractive and will hold securely. When you tap the dowel into place, do not tap it any lower than the surface of the flooring.

If the area to be floored is much larger than usual, you may decide against wood strips. You will find that the wider boards laid in random fashion will meet your taste standards better than the very narrow strips.

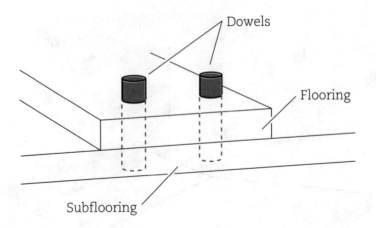

Dowels

Flooring

Subflooring

When one type of flooring meets another, such as tile meeting boards, you can buy a thin wood slat that fits over the joint where the two materials meet. This slat will help convey the impression of a logical flow from one material to the other.

Wood flooring over concrete

You may decide that you would like to install a wood flooring over concrete, perhaps in a basement. You need to install a waterproofing underlayment for the wood surface if the basement floor rests upon earth rather than concrete laid in a suspended fashion.

You can get tubes of mastic and application guns with which to lay rows or wide beads of mastic to which the *sleepers* (supporting timbers) for the floor will be attached. Decide on the layout for the sleepers, usually 16 inches on center, and lay the mastic beads.

Use treated 1-x-2 sleepers (you can rip treated 2-x-4s) and set these into the mastic. You can nail the sleepers to the mastic for better security. Next lay a sheet of polyethylene over the sleepers, and then nail another set of wood strips (also 1-x-2) over the sleepers and attach the flooring to the sleepers.

You can lay blocks of wood by putting down a thin layer of mastic and setting the blocks in the mastic, or you can nail the blocks of wood to the subflooring. Some of the blocks are tongue and groove and can be fitted and nailed easily.

To be sure that your floor is symmetrical, you may need to start in the center of the room and work in all four directions. Measure from one wall to another and mark the midpoint. Then measure from the other two walls and mark that point. In this way you can determine the exact center of the room.

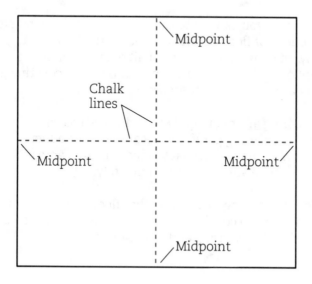

Lay your first block so that the midpoint of the room is at the midpoint of the block. Before laying the block permanently, chalk a line across the room, and as you install each block, see that the edge of the block conforms to the chalk line.

Continue until you have laid blocks in both directions and have one continuous line across the entire room. Then return to the center of the room and lay more blocks in the other directions until you have the room divided into four equal areas.

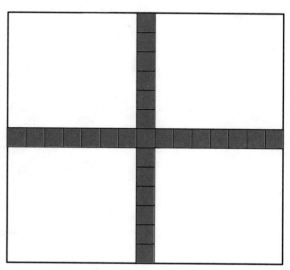

Fill in the areas by laying blocks until each quadrant is covered. Use baseboard molding along the walls.

Kitchens and bathrooms

In kitchen and bathroom areas, you can lay tile squares or continuous tile or linoleum sections. With square tiles, go over the room and knock down any nail heads that protrude even slightly. Fill in all cracks between plywood subflooring panels. When the floor is as smooth as you can get it, chalk off the floor as before and lay the blocks or tiles.

In small areas you can lay bathroom or closet floors by buying large sections of floor covering. Measure the floor area carefully, making special notes of all floor irregularities or interruptions. Measure, mark, and cut the floor covering as carefully as you can. Try it for size several times as you work.

Install flooring after you apply the adhesive coating if any. Position the covering carefully, because once it adheres, it will not move. You may want to lay smooth boards across the floor surface until the adhesive sets completely.

Your dealer can suggest several other floor-covering materials you might want to consider. When this work is done, so is your home, except for painting and touch-up work.

Index